D0866152

Foolproof,

and

Other Mathematical Meditations

Foolproof,

and

Other Mathematical Meditations

Brian Hayes

The MIT Press
Cambridge, Massachusetts
London, England

This book was set in ITC Stone by Brian Hayes. Printed and bound in the United States of America.

Library of Congress Cataloging-in-Publication Data is available.

ISBN: 978-0-262-03686-3

10 9 8 7 6 5 4 3 2 1

To the mathematics community
that has taught me
and charmed me

Contents

Preface

Mathematics is too important and too much fun to be left to the mathematicians. It is an essential tool for understanding the world we live in; as Galileo said, the book of nature is written in mathematical language. At the same time, mathematics is a world apart from the material universe, a realm built not from matter and energy but from pure thought, from the stuff of imagination. It is a universe we can not only explore but also create, inventing new geometries, new kinds of numbers, even new kinds of logic. For example, everyday experience limits us to life in three dimensions, but mathematical spaces have n dimensions—and that innocent-looking symbol n conceals an abundance of possibilities. It might be infinite. It might be a fraction.

Mathematics, in other words, can take you out of this world. Bertrand Russell, who charted the territory and the limits of mathematics a century ago, spoke of opening "windows into a larger and less fretful cosmos."

Who am I to be writing about mathematics, and even trying to *do* a bit of math from time to time? I am not a mathematician—not a native citizen of the Republic of Numbers. But I have been living there, an expatriate litterateur, for most of my adult years. I have struggled to learn the language, immersed myself in the culture and customs, and become an enthusiastic amateur practitioner. My life has been greatly enriched by the experience.

In many quarters, math has acquired a grim reputation: dull, difficult, and detached from the concerns of daily life. More than 2,000 years ago, students were already complaining about the bumpy road they had to negotiate in learning geometry. More recently, a talking Barbie doll opined that "math class is tough." There's no denying that mathematical ideas can be challenging—in some cases even for expert mathematicians. For my part, there are whole subfields of mathematics that are totally opaque to me, where I can't even make sense of the titles of journal articles. But not all math is so forbidding and inaccessible. Some ideas are actually quite simple and quickly mastered, like the little trick for summing a series of consecutive numbers that Carl Friedrich Gauss invented when he was nine or ten years old (see chapter 1). Others are wildly counterintuitive; for example, is that if you put an n-dimensional ball in an n-dimensional box, the ball shrinks away to nothing as n increases (see chapter 10). And there are ideas that require more effort. Understanding the Riemann zeta function—what it is, where it comes from, and why it's important—requires a fairly steep hike up the learning curve. But I would argue that the view from the top is worth it (see chapter 4).

The essays collected in this volume all began with my own struggle to understand some concept or solve a puzzle or trace an event in the history of mathematics. I have described not only the answers that eventually emerged but also the journeys that led me to them. I hope you'll come along. Mathematics may be a foreign country, but it's just the sort of place any adventurous and open-minded traveler will want to visit.

1

Young Gauss Sums It Up

Let me tell you a story, although it's such a well-worn nugget of mathematical lore that you may have heard it already:

> In the 1780s a provincial German schoolmaster gave his class the tedious assignment of summing all the whole numbers from 1 to 100. The teacher's aim was to keep the kids quiet for half an hour, but one young pupil almost instantly produced an answer: $1+2+3+\cdots+98+99+100=5{,}050$. The smart aleck was Carl Friedrich Gauss, who would go on to join the short list of candidates for greatest mathematician ever. Gauss was not a calculating prodigy who added up all those numbers in his head. He had a deeper insight: if you "fold" the series of numbers in the middle and add them in pairs—$1+100$, $2+99$, $3+98$, and so on—all the pairs sum to 101. There are 50 such pairs, and so the grand total is simply 50×101. The more general formula, for a list of consecutive numbers from 1 through n, is $n(n+1)/2$.

The preceding paragraph is my own rendition of this anecdote. I say it's my own, and yet I make no claim of originality. The same tale has been told in much the same way by hundreds of others before me. I've been hearing about Gauss's schoolboy triumph since I was a schoolboy myself.

Although the story is familiar, I had never thought carefully about the events in that long-ago classroom. As I wrote it out in my own words, however, doubts and questions began to nag at me. For example, how did the teacher verify that Gauss's answer was correct? If the schoolmaster already knew the formula for summing an arithmetic series, that would somewhat blunt the drama of the moment. If the teacher *didn't* know, wouldn't he be spending his interlude of peace and quiet doing the same mindless exercise as his pupils?

Soon I was wondering about the provenance and authenticity of the whole story. Where did it come from, and how was it handed down to us? Do scholars take this episode seriously as an event in the life of the mathematician? Or does it belong to the same genre as those stories about Newton and the apple or Archimedes in the bathtub, where literal truth is not the main issue? If we treat the episode as a myth or fable, then what is the moral of the story?

To satisfy my curiosity I began searching libraries and online resources for versions of the Gauss anecdote. Over the course of a few months I collected more than a hundred exemplars, in eight languages. I've discovered others since then, and helpful friends have sent me still more. The sources range from scholarly histories and biographies to textbooks and encyclopedias, and on through children's literature, websites, lesson plans, poems, student papers, YouTube videos, and novels. All the retellings describe what is recognizably the same incident—indeed, I believe they all derive ultimately from a single source—and yet they also exhibit marvelous diversity and creativity, as authors have struggled to fill in gaps, explain motivations, and construct a coherent narrative. (I soon realized that I had done a bit of ad lib embroidery myself.)

After reading all those variations on the story, I still can't answer the factual question, Did it really happen that way? But I think I have learned something about the evolution and transmission of such stories, and about their place in the culture of science and mathematics. I also have some thoughts about how the rest of the kids in the class might have approached their task, and about how the anecdote has been put to use as a lesson for young students of mathematics today.

Wunderkind

I started my survey with five modern biographies of Gauss: books by G. Waldo Dunnington (1955), Tord Hall (1970), Karin Reich (1977), W. K. Bühler (1981), and M. B. W. Tent (2006). The schoolroom incident is related by all of these authors except Bühler. The versions differ in a few details, such as Gauss's age, but they agree on the major points. They all mention the summation of the same series, namely, the integers from 1 to 100, and they all describe Gauss's method in terms of forming pairs that sum to 101.

None of these writers express much skepticism about the anecdote (unless Bühler's silence can be interpreted as doubt). There is no extended discussion of the story's origin or the evidence supporting it. On the other hand, references in some of the biographies did lead me to the key document on which all subsequent accounts seem to depend.

This *locus classicus* of the Gauss schoolroom story is a memorial volume published in 1856, just a year after Gauss's death. The author was Wolfgang Sartorius, Baron von Waltershausen, professor of mineralogy and geology at the University of Göttingen, where Gauss spent his entire academic career. As befits a funerary tribute, it is affectionate and laudatory throughout.

In the portrait Sartorius gives us, Gauss was a *wunderkind*. He taught himself to read, and by age three he was correcting an error in his father's arithmetic. Here is the passage where Sartorius describes Gauss's early schooling in the town of Braunschweig (or Brunswick), near Hanover. The translation, except for two phrases in square brackets, is by Helen Worthington Gauss, a great-granddaughter of the mathematician.

> In 1784 after his seventh birthday the little fellow entered the public school where elementary subjects were taught and which was then under a man named Büttner. It was a drab, low school-room with a worn, uneven floor. . . . Here among some hundred pupils Büttner went back and forth, in his hand the switch which was then accepted by everyone as the final argument of the teacher. As occasion warranted he used it. In this school—which seems to have followed very much the pattern of the Middle Ages—the young Gauss remained two years without special incident. By that time he had reached the arithmetic class in which most boys remained up to their fifteenth year.
>
> Here occurred an incident which he often related in old age with amusement and relish. In this class the pupil who first finished his example in arithmetic was to place his slate in the middle of a large table. On top of this the second placed his slate and so on. The young Gauss had just entered the class when Büttner gave out for a problem [the summing of an arithmetic series]. The problem was barely stated before Gauss threw his slate on the table with the words (in the low Braunschweig dialect): "There it lies." While the other pupils busily continued [calculating, multiplying and adding], Büttner, with conscious dignity, walked back and forth, occasionally throwing an ironical, pitying glance toward this the youngest of the pupils. The boy sat quietly with his task ended, as fully aware as he always was on finishing a task that the problem had been correctly solved and that there could be no other result.
>
> At the end of the hour the slates were turned bottom up. That of the young Gauss with one solitary figure lay on top. When Büttner read out the answer, to the surprise of all present that of young Gauss was found to be correct, whereas many of the others were wrong.

Incidental details from this account reappear over and over in later tellings of the story. The ritual of piling up the slates is one such feature. (It must have been quite a teetering heap by the time the hundredth slate was added!) Büttner's switch (or cane, or whip) also made frequent appearances until the 1970s but is less common now; we have grown squeamish about mentioning such barbarities.

What's most remarkable about the Sartorius telling of the story is not what's there but what's absent. There is no mention of the numbers from 1 to 100, or any other specific arithmetic progression. And there is no hint of the trick or technique that Gauss invented to solve the problem; the idea of combining the numbers in pairs is not discussed, nor is the general formula for summing a series. Perhaps Sartorius thought the procedure was so obvious it needed no explanation. Or maybe he didn't know the secret himself.

A word about the bracketed phrases. Strange to report, the Worthington Gauss translation *does* mention the first 100 integers. Where Sartorius writes simply "einer arithmetischen Reihe" ("an arithmetic series"), Worthington Gauss inserts "a series of numbers from 1 to 100." I cannot account for this interpolation. I can only guess that Worthington Gauss, under the influence of later works that discuss the 1–100 example, was trying to help out Sartorius by filling in an omission.

The second bracketed passage marks an elision in the translation. Where Sartorius has the pupils "rechnen, multiplizieren und addieren" ("calculating, multiplying and adding"), Worthington Gauss writes just "adding." I'll have more to say on this point later.

Making History

If Sartorius did not specify a series running from 1 to 100, where did those numbers come from? Could there be some other document from Gauss's era that supplies the missing details? Perhaps on one of those occasions when Gauss told the story "with amusement and relish," someone left a record of the event. The existence of such a corroborating document cannot be ruled out, but at present there is no evidence for it. None of the works I have seen makes any allusion to another early source. If an account from Gauss's lifetime exists, it remains so obscure that it can't have had much influence on other tellers of the tale.

Among all the published versions of the story I've been able to track down (see figure 1.1), the earliest that mentions a specific series of numbers is a speech written in the 1870s by Hans Sommer, then the director of the technical college in Braunschweig. Sommer's account is also the first to explain the trick for quickly summing the numbers. The series he cites is not 1–100 but rather 1–40, and he suggests it as merely an example of what the students' task might have been like. Here is the crucial paragraph (translated from the German):

> Starting in 1784 Gauss attended the St. Catherine public school, which was under the direction of a certain Büttner. When Gauss entered the arithmetic class two years later, the children were given the problem of adding up a series of successive numbers, for instance 1 to 40. When each student had finished his calculation, he had to put his slate on the classroom table. Gauss, after a short reflection, wrote the result, and threw it on the table with the words, "There it lies!" Other pupils went through the most arduous labors and finished long afterwards. Contrary to Büttner's expectation, Gauss's result was exactly right. The ingenious boy had immediately noticed that 1 and 40 are the same as 2 and 39, or 3 and 38, or any two numbers equally distant from the beginning and the end of the series. Each such pair sums to 41. There were 20 pairs, so the total had to be 20 × 41, or 820. Thus the nine-year-old Gauss had at first sight recognized and applied the summation principle for arithmetic series.

Sommer was a Göttingen-educated mathematician and physicist, although he is better known today as a composer and musician, and as a friend of Richard and Cosima Wagner. Apparently he delivered this speech in Braunschweig on April 30, 1877, the 100th anniversary of Gauss's birth. However, the text was not published until 1894, and it appeared then in a strange and sinister context—a volume titled *Contributions to the German Jewish Question.* (Gauss was being celebrated as an exemplar of non-Jewish German culture.)

Sommer's version of the story, including his 1–40 example, was taken up by others over the next few decades. In 1899 it was retold by Paul Möbius, a nephew of August Möbius, the discoverer of the Möbius band. Curiously, Möbius cites a different Gauss biography, Friedrich Winnecke's *Life and Work* of 1877, but his source for the schoolroom scene is clearly Sommer. The wording of the two passages is nearly identical, including the 1–40 example and the pairing idea.

In 1915 Wilhelm Ahrens included Gauss's tale of schoolboy triumph in a collection of *Mathematical Anecdotes*. Ahrens cites no sources, but his text appears to be based on the Sommer (or Möbius) version. It presents the 1–40 example and gives a similar but slightly more elaborate explanation of the pairing method.

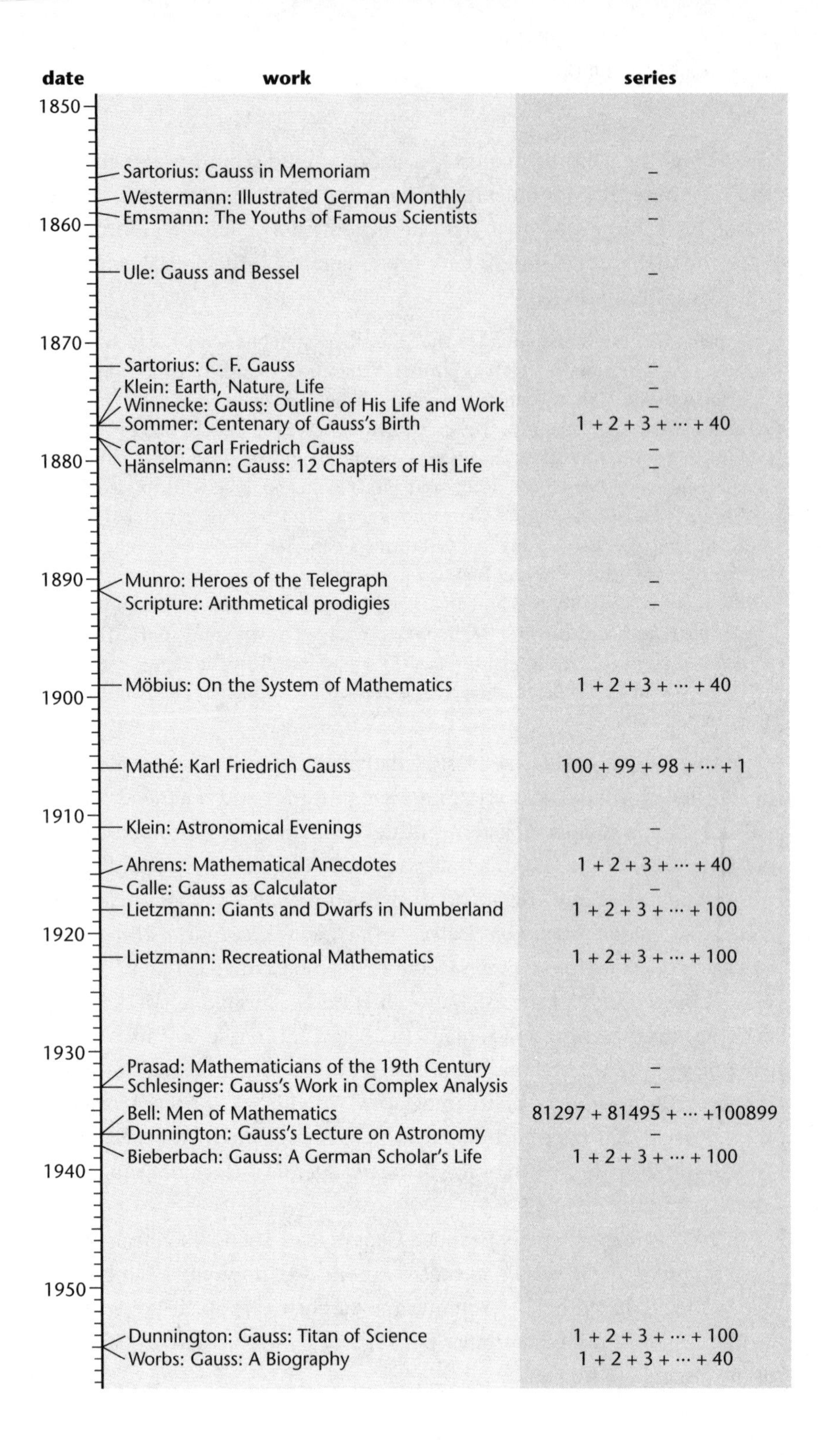
date
work
series
1850
1860
1870
1880
1890
1900
1910
1920
1930
1940
1950
Sartorius: Gauss in Memoriam
–
Westermann: Illustrated German Monthly
–
Emsmann: The Youths of Famous Scientists
–
Ule: Gauss and Bessel
–
Sartorius: C. F. Gauss
–
Klein: Earth, Nature, Life
–
Winnecke: Gauss: Outline of His Life and Work
–
Sommer: Centenary of Gauss's Birth
1 + 2 + 3 + ··· + 40
Cantor: Carl Friedrich Gauss
–
Hänselmann: Gauss: 12 Chapters of His Life
–
Munro: Heroes of the Telegraph
–
Scripture: Arithmetical prodigies
–
Möbius: On the System of Mathematics
1 + 2 + 3 + ··· + 40
Mathé: Karl Friedrich Gauss
100 + 99 + 98 + ··· + 1
Klein: Astronomical Evenings
–
Ahrens: Mathematical Anecdotes
1 + 2 + 3 + ··· + 40
Galle: Gauss as Calculator
–
Lietzmann: Giants and Dwarfs in Numberland
1 + 2 + 3 + ··· + 100
Lietzmann: Recreational Mathematics
1 + 2 + 3 + ··· + 100
Prasad: Mathematicians of the 19th Century
–
Schlesinger: Gauss's Work in Complex Analysis
–
Bell: Men of Mathematics
81297 + 81495 + ··· +100899
Dunnington: Gauss's Lecture on Astronomy
–
Bieberbach: Gauss: A German Scholar's Life
1 + 2 + 3 + ··· + 100
Dunnington: Gauss: Titan of Science
1 + 2 + 3 + ··· + 100
Worbs: Gauss: A Biography
1 + 2 + 3 + ··· + 40

Thus by the early twentieth century at least three published accounts had suggested that the problem Büttner gave his students was summing the series 1 through 40. But then the story took a strange twist. In 1918 Walther Lietzmann, a Göttingen authority on the teaching of mathematics, published *Giants and Dwarfs in Numberland,* a book for children and math enthusiasts. In telling the Gauss schoolroom story, Lietzmann cites Ahrens as his source, but without explanation he substitutes 1–100 for 1–40. As far as I know, this is the earliest published version of the story in which young Gauss sums up the numbers from 1 to 100. Lietzmann repeated the story, with the same details, in a 1922 book.

There was one earlier near miss. In 1906 Franz Mathé, a teacher at the technical college in Reichenberg (now the Czech city of Liberec), published a brief biography of Gauss in which Büttner's assignment is to "add all the numbers from 100 down to 1," the reverse of the usual sequence. No one in the subsequent history of the tale seems to have followed that suggestion.

The 1–100 example appears again in 1938, in a biography of Gauss by Ludwig Bieberbach (a mathematician notorious as the principal instrument of Nazi anti-Semitism in the German mathematical community). This may well be a case of independent invention, since Bieberbach's wording has no clear echoes of Lietzmann (or any other prior author except Sartorius, whose work he cites).

In 1955 (the centenary of Gauss's death), two more book-length biographies appeared. Erich Worbs, in *Carl Friedrich Gauss: Ein Lebensbild,* stayed with the 1–40 example first introduced by Hans Sommer almost 75 years earlier. G. Waldo Dunnington, in *Carl Friedrich Gauss: Titan of Science,* chose 1–100. Dunnington is an interesting case. For the most part, his telling of the story is nearly a word-for-word translation of Sartorius, but then in the middle of it he interjects the 1–100 summation, without any hint that this is merely an illustrative example, not an established biographical fact.

Figure 1.1 A timeline of the Gauss schoolroom anecdote records about 25 tellings of the tale published in the 100 years after Gauss's death in 1855. The first account, by Wolfgang Sartorius von Waltershausen, seems to be the ultimate source for all the others. Most early versions mentioned no specific series of numbers and gave no hint of how Gauss solved the problem. Hans Sommer suggested the example series 1–40 in an 1877 address (not published until 1894); Walther Lietzmann was apparently the first to propose 1–100 in 1918.

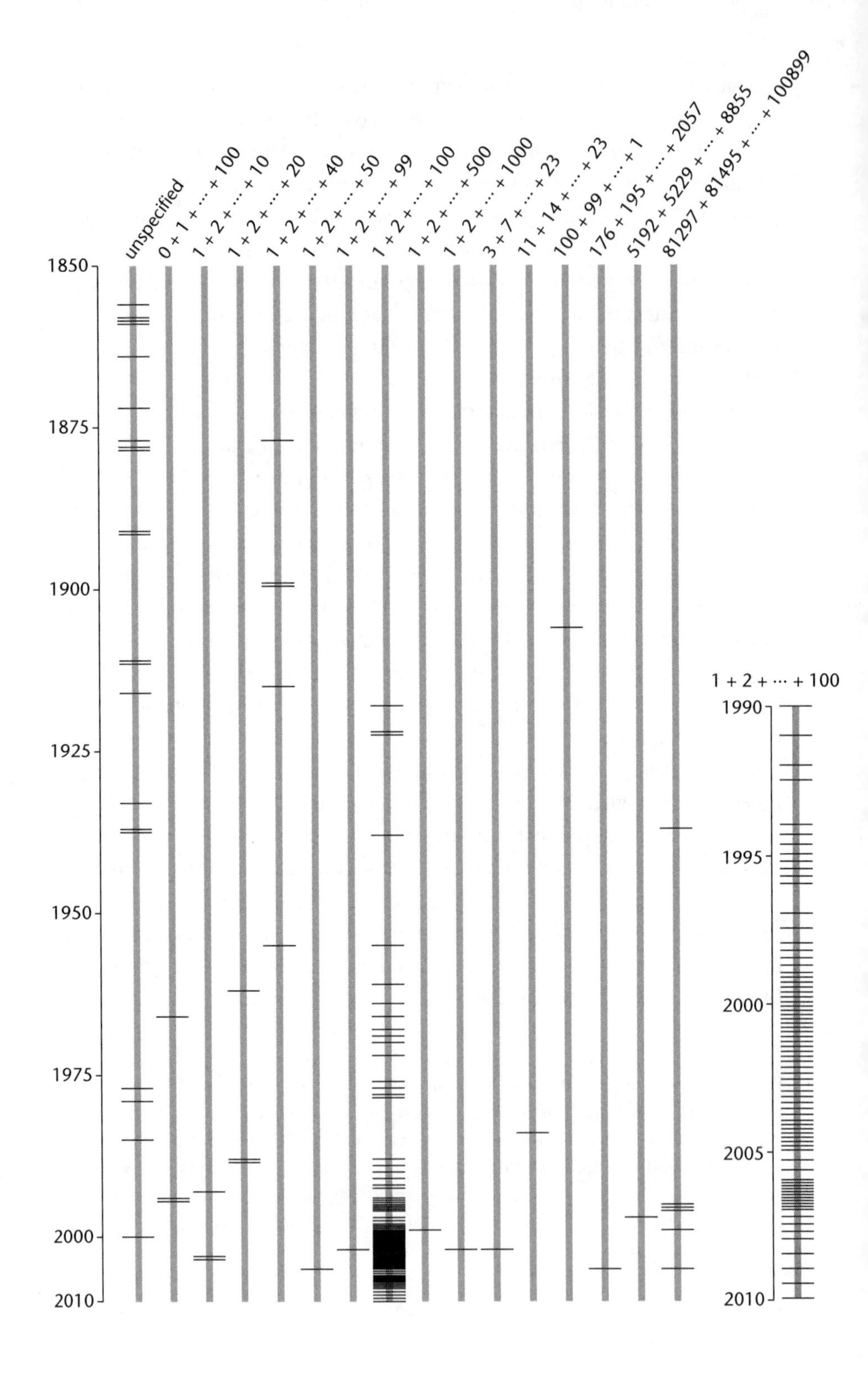

unspecified
0 + 1 + ··· + 100
1 + 2 + ··· + 10
1 + 2 + ··· + 20
1 + 2 + ··· + 40
1 + 2 + ··· + 50
1 + 2 + ··· + 99
1 + 2 + ··· + 100
1 + 2 + ··· + 500
1 + 2 + ··· + 1000
3 + 7 + ··· + 23
11 + 14 + ··· + 23
100 + 99 + ··· + 1
176 + 195 + ··· + 2057
5192 + 5229 + ··· + 8855
81297 + 81495 + ··· + 100899
1850
1875
1900
1925
1950
1975
2000
2010
1 + 2 + ··· + 100
1990
1995
2000
2005
2010

Dunnington had written two earlier essays on Gauss. In 1927 he did not mention the Büttner episode at all; in 1937 he told the story but gave no specific series of numbers.

One more telling of the story deserves to be mentioned here, because it is probably the most widely read version. Eric Temple Bell included the anecdote in his *Men of Mathematics*, first published in 1937. The same essay was reprinted in James R. Newman's *World of Mathematics* 20 years later. Both works are still in print. Bell has a reputation as a highly inventive writer (a trait not always considered a virtue in a biographer or historian). He turns the Braunschweig schoolhouse into a scene of gothic horror: "a squalid relic of the Middle Ages run by a virile brute, one Büttner, whose idea of teaching the hundred or so boys in his charge was to thrash them into such a state of terrified stupidity that they forgot their own names." Very cinematic! When it comes to the arithmetic, he chooses a more formidable problem than other authors, but he is also careful to distinguish between fact and conjecture. He doesn't claim to know the actual series given in the class assignment. He writes, "The problem was of the following sort, $81297 + 81495 + 81693 + \cdots + 100899$, where the step from one number to the next is the same all along (here 198), and a given number of terms (here 100) are to be added." (Personally, I'd have a hard time even writing that problem on a small slate, much less solving it.)

In the years since 1950, the 1–100 series has won the popularity contest by a wide margin (see figure 1.2). Of 130 examples that I've recorded in that period, all but 25 describe the problem as summing the consecutive integers from 1 to 100. Yet other variants are not dying out. Some differ only slightly: 0–100 or 1–99. Several authors seem to feel that adding up 100 numbers is too big a job for primary school students, and so they trim the scope of the assignment, suggesting 1–80, or 1–50, or 1–20, even 1–10. A few others apparently think that 1–100 is too easy, and so they give 1–500 or 1–1,000, or else they propose a series in which the difference between

Figure 1.2 Variations in the Gauss story are plotted over a period of 150 years. Each column shows the dates of tellings that cite a single exemplar series (or no series at all in the case of the leftmost column). Most of the early versions left the series unspecified. The first series to be mentioned was 1–40, then 1–100 became common after 1950 and dominant after 1990 *(see expanded view of this series at far right)*. Yet other variants continue to appear sporadically.

successive terms is a constant other than 1, such as the sequence 3, 7, 11, 15, 19, 23, 27.

The Narrative Urge

It's a challenge to sort out patterns of influence and transmission in such a collection of stories. When a later author mentions the series $81297 + 81495 + \cdots$, we can be pretty sure those numbers came from Eric Temple Bell. When the example given is 1–100, however, it's not so easy to trace the line of inheritance—if there is one. And the dozen or so other sequences that appear in the literature argue for a high rate of mutation; every one of those examples had to be invented at least once.

Tellers of a tale like this one seem to work under a special dispensation from the usual rules of history writing. Authors who would not dare to alter a fact such as Gauss's place of birth or details of his mathematical proofs don't hesitate to embellish this anecdote, just to make it a better story. They pick and choose from the materials available to them, taking what they need and leaving the rest—and if nothing at hand suits the purpose, then they invent! For example, several authors show a familiarity with Bell's version of the story, quoting or borrowing distinctive phrases from it ("a virile brute"), but they decline to go along with Bell's choice of a series beginning with 81297, falling back instead on the old reliable 1–100 or inserting something else entirely. Thus it appears that what is driving the evolution of this story is not just the accumulation of errors of transmission, as in the children's game "whisper down the lane"; authors are deliberately choosing to "improve" the story, to make it a better narrative.

For the most part, I would not criticize this practice. Effective storytelling is surely a legitimate goal, and outside of formal scholarly works, a bit of embroidery on the bare fabric of the plot does no harm. A case in point is the theme of busywork found in most recent tellings of the story (including my own). It seems we feel a need to explain why Büttner would give his pupils such a long and dreary exercise. But Sartorius says nothing at all about Büttner's motivation, nor do any of the other nineteenth-century works I've consulted. That he wanted to keep the kids quiet while he took a break is a modern interpretation. It's probably wrong—at best it's unattested—and yet it answers a need of readers today.

In the same spirit, many authors confront one of the questions that got me started on this quest. How did Büttner do the math? Bell is adamant that Büttner knew the formula beforehand; others say he learned the trick only when Gauss explained it to him. An example of the latter position is the following account written in 2001 by three fifth-grade students, Ryan, Jordan, and Matthew:

> When Gauss was in elementary school his teacher Master Büttner did not really like math so he did not spend a lot of time on the subject. One of the problems his teacher gave the class was "add all the whole numbers from 1 to 100." His teacher Master Büttner was amazed that Gauss could add all the whole numbers 1 to 100 in his head. Master Büttner didn't believe Gauss could do it, so he made him show the class how he did it. Gauss showed Master Büttner how to do it and Master Büttner was amazed at what Gauss just did.

On the question of historical plausibility, I have to side with Bell. A schoolmaster in Büttner's era would have known the series summation formula. But if we choose to see the tale not as history but as inspirational literature—as a fable about quick-witted youth sweeping away the tired habits of an older generation—then Ryan, Jordan, and Matthew surely have it right.

Although Gauss may well have discovered the summation method entirely on his own, he was certainly not the first to do so. Leonardo of Pisa, or Fibonacci, knew the trick more than 500 years earlier. In his *Liber Abaci* he wrote (in a translation by L. E. Sigler),

> When you wish to sum a given series of numbers which increases by some given number, as increasing by ones, or twos, or threes, or any other numbers, then you multiply half the number of numbers in the series times the sum of the first and last numbers in the series, or you multiply half the sum of the first and last numbers in the series by the number of numbers in the series, and you will have the proposition.

Earlier still, in the eighth century, the algorithm turns up among 53 "Problems to Sharpen Youths," compiled by Alcuin of York, an English scholar working in the court of Charlemagne. Alcuin describes a ladder with 100 steps, with one dove on the first step, two doves on the second step, and so on. Thus he had already hit upon on the specific 1–100 example that is now so popular. His solution method is somewhat peculiar. He forms the 49 pairs $1+99$, $2+98$, ..., $49+51$, where each pair yields 100; then, to the product $49 \times 100 = 4900$ he adds the unpaired numbers 100 and 50.

And there's more! Even in Alcuin's time, the problem was already ancient. Archimedes derived the summation formula in his work *On Spirals*, circa 225 B.C.E.

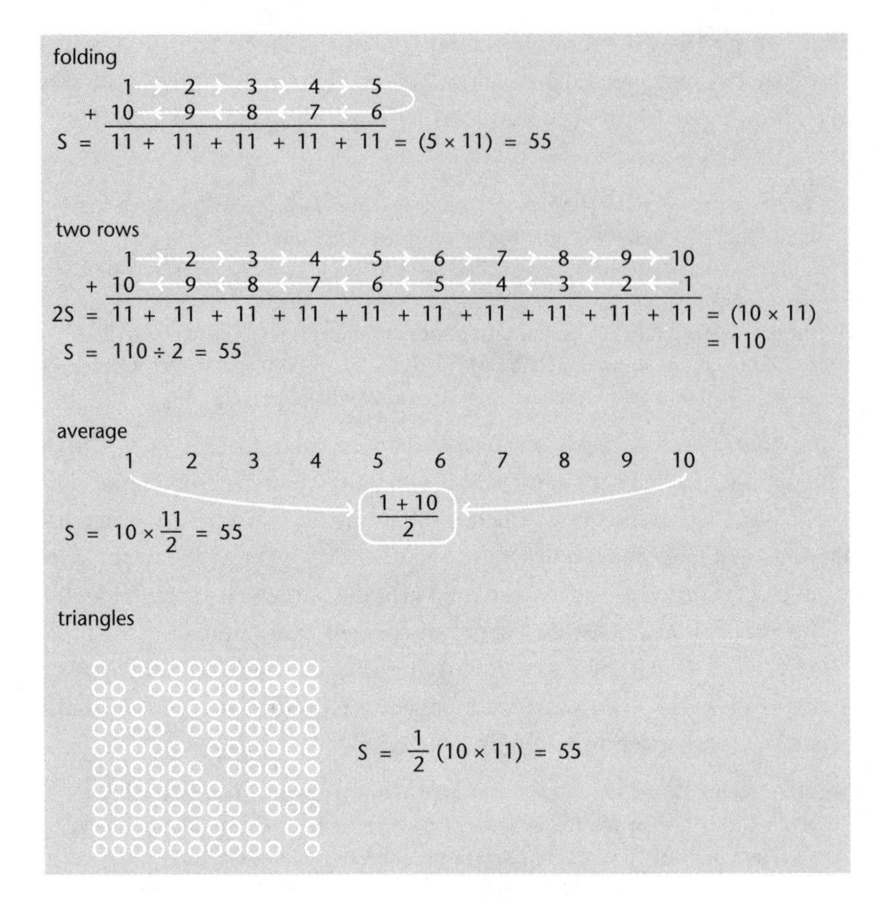

Figure 1.3 How did young Gauss sum the arithmetic series? He probably observed that pairs of elements from opposite ends of the series all have the same sum. One algorithm folds the series in half with a hairpin bend. Another writes the series twice, once in ascending and once in descending order. A third method calculates the average from the first and last elements. Finally, the formula for summing the first n natural numbers also generates the nth triangular number; the sum is half the area of an $n \times (n+1)$ rectangle.

Summing Up

Just as we can't know what specific series Gauss was asked to sum, we can only guess about the details of how he solved the problem (see figure 1.3). The algorithm that I suggested—folding the sequence in half, then adding the first and last elements, the second and next-to-last, and so on—is not the only possibility. A related but subtly different algorithm is mentioned by many authors. The idea is to write down the series twice, once forward and once backward, and then add corresponding elements. For the familiar series 1–100 this procedure yields 100 pairs of 101, for a total of 10,100; then, since the original series was duplicated, we need to divide by 2, arriving at the correct answer 5,050. The advantage of this scheme is that it works the same whether the length of the sequence is odd or even, whereas the folding algorithm requires some fussy adjustments to deal with an odd-length series.

A third approach to the summation problem strikes me as better still. The root idea is that for any finite set of numbers, whether or not the numbers form an arithmetic progression, the sum is equal to the average of all the elements multiplied by the number of elements. Thus if you know the average, you can easily find the sum. For most sets of numbers, this fact is not very useful, because the only way to calculate the average is first to calculate the sum and then divide by the number of elements. For an arithmetic progression, however, there is a shortcut. The average over the entire series is equal to the average of the first and last elements (or the average of any other elements symmetrically arrayed around the midpoint). If this was Gauss's secret weapon, then his mental multiplication was not 50×101 but $100 \times 50\frac{1}{2}$.

All three of these ideas—and a few more besides—have been presented by one author or another as the method that Gauss discovered during his first arithmetic lesson. Expressed as formulas for summing consecutive integers from 1 through n, the three rules (folding, double rows, average) look like this:

$$\frac{n}{2}(n+1), \quad \frac{n(n+1)}{2}, \quad n\frac{(n+1)}{2}.$$

Mathematically, they are equivalent; for the same value of n, they all produce the same answer. But the computational details are different and, more important, so are the reasoning processes that lead to these formulas.

There is yet another way of thinking about the summation process: $n(n+1)/2$ has been known since antiquity as the formula for the triangular numbers, those in the sequence 1, 3, 6, 10, 15, 21,... (To see what's triangular about them, think of the standard arrangement of 10 bowling pins or 15 billiard balls.) Thus some authors suggest that Gauss solved the problem geometrically, forming an $n \times (n+1)$ rectangle and cutting it in half along the diagonal.

Doing It the Hard Way

So much for how the prodigious Carl Friedrich Gauss solved the problem. What about the rest of the students in the class? I invite you to take a sheet of paper and actually try adding the numbers from 1 to 100. The process may not be quite as lengthy and tedious as you imagine.

What I discovered when I tried this experiment is that it's really hard to stick to doing it the hard way. You may set out to plod dutifully through all 99 addition operations, but shortcuts present themselves even when you're not looking for them. Suppose you adopt the standard primary school method, writing down all 100 numbers in a tall stack and then starting work on the units column. After the first ten digits in that column, the partial sum is 45. Then you observe that the next ten digits are identical to the first ten and so must also add up to 45, and likewise the next ten; the pattern continues all the way down the units column. Beginning work on the tens column, you find ten 1s followed by ten 2s, then ten 3s, and so on. Surely any student who notices these patterns would not slavishly add all the repeated numbers one by one.

On a small slate or a sheet of paper, it's difficult to write 100 numbers in one column, and so students might break the task down into subproblems. Suppose you start by adding the numbers from 1 to 10, for a sum of 55. Then you discover that the sum of 11 through 20 works out to 155, and 21 through 30 yields 255. Again, how far would you continue before spotting the trend and taking a shortcut?

Consider again how Sartorius described the work of Gauss's classmates. They were busily "rechnen, multiplizieren und addieren," a phrase that I initially translated as "counting, multiplying, and adding." Andreas M. Hinz of Ludwig-Maximilians University in Munich suggests instead "calculating, multiplying and adding," where the latter two terms are parenthetical, explaining what activities go into "calculating." In any case, it's clear that Sartorius believed there was more going on than mere "adding," the only term that survived in the translation by Helen Worthington Gauss.

Admittedly, such shortcuts can't match the elegance and ingenuity of Gauss's method. They are tied to the decimal representation of numbers, and they also don't generalize as well to arithmetic progressions other than lists of consecutive integers. But they do remind us that there's usually more than one good way to solve a problem.

I suspect that only one kind of student would ever be likely to add the numbers from 1 through 100 by performing 99 successive additions—namely, a student using a computer or a programmable calculator. And for that student, the clever Gaussian trick is not necessarily the best answer.

We can hope that a modern Büttner—deprived of his whip, of course, and teaching in a classroom where computers have replaced slates—would not be drilling students on skills of such dubious utility as adding up a long column of numbers by hand. But the new Büttner might well ask her pupils to write a program to calculate the sum of the integers from 1 through *n*. A twenty-first-century Gauss might come up with something like this:

```
define clever_sum(n)
    return n*(n+1)/2
```

Then, of course, she would fling her laptop on the table and cry, "There it lies!" The program is concise, clear, and highly efficient. Its running time is essentially the same no matter how large *n* becomes (at least until the numbers exceed the size of a machine register).

Other students would doubtless stick closer to the statement of the problem, writing a `for` loop that steps through the integers from 1 through *n* and adds each of them to a running total:

```
define plodding_sum(n)
    total = 0
    for i from 1 to n
        total = total + i
    return total
```

This approach is more verbose, and the running time increases in proportion to the value of *n*, making it slow for very large *n*. But one shouldn't be too quick to dismiss the plodding solution.

Suppose Büttner assigns a new task: Write a program that returns the sum of the squares of all the integers from 1 to *n*. For the plodders this is easy—just replace the line `total = total+i` with `total = total+(i*i)`. What about young Gauss? It turns out there *is* a formula for the sum of the squares of consecutive integers. She could replace `n*(n+1)/2` with `n*(n+1)*((2*n)+1)/6`. But even Gauss might need to do a bit of work to discover that formula or to explain why it gives correct results.

Now Büttner asks the class to sum up the reciprocals of a sequence of whole numbers,

$$\frac{1}{1}+\frac{1}{2}+\frac{1}{3}+\cdots+\frac{1}{n}.$$

Again the plodders need only a moment to revise their programs, inserting the line `total = total+(1/i)`. But Gauss is stymied. There is no closed-form formula that yields the exact value of this series. Not every problem in mathematics has an elegant shortcut that yields answers effortlessly. (This is something the adult Gauss knew very well; he filled many notebooks with the results of laborious calculations.)

The Moral of the Fable

The story of Gauss and his conquest of the arithmetic series has a natural appeal to young people. After all, the hero is a child—a child who outwits a "virile brute." For many students, that is surely an inspiration. But I worry a little that the constant repetition of stories like this one may leave the impression that mathematics is a game suited only to those who go through life continually throwing off sparks of brilliance.

On first hearing this fable, most students surely want to imagine themselves in the role of Gauss. Sooner or later, however, most of us discover we are one of the less distinguished classmates; if we eventually get the right answer, it's by hard work rather than native genius. I would hope that the story could be told in a way that encourages those students to keep going. And perhaps it can be balanced by other stories showing there's a place in mathematics for more than one kind of mind.

As a step in that direction, here's another famous mathematical anecdote: the story of von Neumann and the fly. I'm going to tell it from memory, meaning that I'll doubtless introduce a few gratuitous variants like those that have enriched the literature of Gauss's boyhood.

In the early 1950s the following problem was making the rounds. Two trains are 20 miles apart on the same track, coming together at 10 miles per hour each. A fly takes off from the nose of one locomotive, flies at 20 miles per hour to the nose of the other locomotive, then instantly turns around and continues shuttling back and forth between the two trains until they smash together. How many miles does the fly fly?

The problem looks messy. It seems you'll need to trace the fly's zigzag trajectory and figure out the positions of the trains whenever the fly reverses course. Worse, if the fly is infinitely small, it turns around infinitely many times before it gets smashed between the locomotives. But there's an easier way. Note that the trains will collide after exactly one hour, and the fly will be moving at 20 miles per hour the whole time.

One day a friend presented the problem to John von Neumann, a mathematician, physicist, and computer scientist who was legendary for his quick mind and problem-solving skills. Von Neumann listened, paused for a few seconds, and then gave the correct answer: 20 miles.

"Ah, you already knew the trick," said his friend.

"What trick?" von Neumann replied. "I just summed the infinite series."

Acknowledgments

I first began collecting versions of the Gauss anecdote in 2005. In those days few books were available online, so I prowled the stacks of libraries, with generous assistance from several librarians. Friends also helped with tracking down obscure volumes, and with translations. I particularly want to thank Johannes Berg of the University of Cologne; Sally Bosken of the U.S. Naval Observatory; Caroline Grey of the Johns Hopkins University libraries; Stephan Mertens of the University of Magdeburg; Ivo Schneider of the Bundeswehr University, Munich; Margaret Tent of the Altamont School in Birmingham, Alabama; and Mary Linn Wernet of the Northwestern State University libraries in Natchitoches, Louisiana.

After the first version of this essay was published in 2006, I received a flood of further contributions from friends and interested readers. Barry Cipra turned up more than two dozen new examples, both in printed books and in online resources. Herb Acree, skillfully searching Google Books, provided leads to several crucial nineteenth-century sources, leading to a major revision in my understanding of the early history of the anecdote. James Grant pointed out Fibonacci's place in the history of the problem, and Donald E. Knuth brought Aristotle's treatment of series summation to my attention. Andreas M. Hinz corrected several errors in my interpretation of German texts, as well as an arithmetic error that had gone uncaught for a decade. Other valuable contributions and commentary came from Mark Auslander, Umberto Barreto, Said Boutiche, Robert Dickey, Hans Magnus

Enzensberger, James Grant, Colm Mulcahey, Henry Picciotto, and Christian Siebeneicher.

Further material on the Büttner-Gauss anecdote is posted on the web at http://bit-player.org/gauss-links. Included are a complete bibliography, a chronological table, and excerpts from more than 150 tellings of the story in the original languages.

2

Outside the Law of Averages

You've probably heard of Lake Wobegon, the little town in Minnesota where all the children are above average. There's been much head-scratching about this statistical miracle. What happens to the kids who fail to surpass themselves? Are they shipped across the lake to another town, where all the children are *below* average? That practice wouldn't necessarily work to the detriment of either community. It might be like the migration from Oklahoma to California during the Dust Bowl years, which Will Rogers said raised the average intelligence of both states.

One small town that beats the law of averages is strange enough, but even more mystifying is the finding that Lake Wobegon is not unique—that in fact *everyone* is above average. In 1987 John J. Cannell, a West Virginia physician and activist, discovered that all 50 states reported that their children do better than the national average on standardized tests.

I can't promise to resolve these paradoxes. On the contrary, I'm going to make matters worse by describing still more funny business in the world of averages. The story that follows is about a data distribution that simply has no average. Given any finite sample drawn from the distribution, you are welcome to apply the usual algorithm for the arithmetic mean—add up the values and divide by the size of the sample—but the result won't mean much. Whatever average you calculate in this way, you can improve it just by taking a bigger sample. Perhaps this is the secret of the Lake Wobegon school board.

The existence of such better-than-average averages is not a new discovery; the phenomenon was already well known a century ago, and distributions with this property have become a hot topic in the past decade. Recently I stumbled upon a particularly simple illustration of the concept, and that's the story I tell here.

Facts about Factorials

It all begins with the factorial function, a familiar item of furniture in several areas of mathematics, including combinatorics and probability theory. The factorial of a positive whole number n is the product of all the integers from 1 through n inclusive. For example, the factorial of 6 is $1 \times 2 \times 3 \times 4 \times 5 \times 6 = 720$.

The standard notation for the factorial of n is $n!$. This use of the exclamation point was introduced in 1808 by Christian Kramp, a mathematician from Strasbourg. Not everyone was enthusiastic about it. Augustus De Morgan, an eminent British mathematician and logician, complained in 1842 that the exclamation points give "the appearance of expressing surprise and admiration that 2, 3, 4, &c. should be found in mathematical results."

One common application of the factorial function is in counting permutations, or rearrangements of things. If six people are sitting down to dinner, the number of ways they can arrange themselves at the table is 6!. It's easy to see why. The first person can choose any of the six chairs, the next person has five places available, and so on until the sixth diner is forced to take whatever seat remains.

The factorial function is notorious for its rapid rate of growth: 10! is already in the millions, and 100! is a number with 158 decimal digits. As n increases, $n!$ grows faster than any polynomial function of n, such as n^2 or n^3, or any simple exponential function, such as 2^n or e^n. Indeed you can choose any constant k, and make it as large as you please, and there will still be some value of n beyond which $n!$ exceeds both n^k and k^n. (On the other hand, $n!$ grows slower than n^n.)

The steep increase in the magnitude of $n!$ becomes an awkward annoyance when you want to explore factorials computationally. A programming language that packs integers into 32 binary digits cannot reach beyond 12!, and even 64-bit arithmetic runs out of room at 20!. To go further requires a language or a program library capable of handling arbitrarily large integers.

In spite of this inconvenience, the factorial function is an old favorite in computer science. Often it is the first example mentioned when introducing the concept of recursion, as in this procedure definition:

```
define f!(n)
  if n=1
    then return 1
    else return n*f!(n-1)
```

One way to understand this definition is to put yourself in the place of the procedure. You are the factorial oracle, and when someone gives you an n, you must respond with $n!$. Your task is easy if n happens to be 1, since calculating 1! doesn't take much effort. If n is greater than 1, you may not know the answer directly, but you *do* know how to find it: just get the factorial of $n-1$ and then multiply the result by n. Where do you find the factorial of $n-1$? Simple. Ask yourself—you're the oracle!

This self-referential style of thinking is something of an acquired taste. For those who prefer loops to recursions, here is another definition of the factorial:

```
define f!(n)
  product ← 1
  for x in n downto 1
    product ← product * x
  return product
```

In this case it's made explicit that we are counting down from n to 1, multiplying as we go. Of course, we could just as easily count *up* from 1 to n; the result would be the same. Indeed, we could arrange the n numbers in any of $n!$ permutations. All the arrangements are mathematically equivalent, although some ways of organizing the computation are more efficient than others.

Factoidals

Playing at the keyboard one day, I came up with the following factorial-like procedure:

```
define f?(n)
  r ← random(1, n)
    if r = 1
      then return 1
      else return r * f?(n)
```

This might be a buggy version of a program intended to calculate factorials, but it actually does something a good deal stranger. The auxiliary procedure `random(1, n)`, invoked in the second line, is assumed to return an integer selected at random from the range 1 through n. Thus the program chooses a series of random integers, multiplies them, and stops when `random(1, n)` happens to yield a 1. It's like rolling an n-sided die and keeping a running product of all the numbers seen until a 1 appears.

Here are the results of a few sample runs of the program, for $n=7$:

$$4\times3\times2\times4\times7\times2\times1=1{,}344,$$

$$2\times5\times4\times5\times4\times3\times6\times2\times2\times5\times1=288{,}000,$$

$$7\times5\times5\times1=175,$$

$$4\times7\times6\times2\times6\times5\times6\times3\times5\times3\times3\times3\times7\times4\times3\times3\times7\times2\times5\times1=432{,}081{,}216{,}000.$$

Note that 7! is equal to 5,040, a value that doesn't turn up in this sample. The results that do appear are scattered over quite a wide range.

This randomized analogue of the factorial function needs a name. I shall call it the factoidal function. Also, risking the ire and ridicule of some latter-day De Morgan, I adopt the notation $n?$ to suggest the nondeterministic nature of the computation.

Strictly speaking, the factoidal function isn't a function—not in the mathematical sense of that word. An essential property of a function is that applying it to the same argument always yields the same value. The function call `f!(7)` will produce the value 5,040 time after time. But, as we have just seen, `f?(7)` is likely to return a different value every time it is invoked, depending on the vagaries of the random number generator. For that matter, the procedure might not return any value at all; the program could run forever. The `f!` procedure is guaranteed to terminate, because on each recursive call (or each passage through the loop) the multiplier gets smaller, and eventually it has to reach 1. But `f?` halts only when the roll of a die comes up 1, which might never happen.

In practice, the program always terminates, one way or another. A back-of-the-envelope calculation shows there's about a two-thirds chance that `random(1, n)` will produce a 1 in n trials. And a 1 is almost certain to turn up within $5n$ trials. Thus if n is small (less than 1,000, say), `f?(n)` will almost surely succeed in calculating a value. If n is very large, the product being computed is likely to fill up all available memory, and the program will terminate with an error message.

Because of the element of randomness in the factoidal definition, it makes no sense to ask about *the* value of $n?$. The best we can hope for is to understand the statistical distribution of values. As a rule, this would mean estimating quantities such as the average value and the variance or standard deviation. But those familiar statistical tools are problematic in this case, so let's start by asking an easier question: How do factoidals compare with factorials? Is $n?$ usually larger than $n!$, or smaller? (They can also be equal, of course, but as n increases, the probability of such an exact match goes to zero.)

When I first began puzzling over this question, I convinced myself that $n?$ would usually exceed $n!$. Mercifully, I have forgotten the details of the "reasoning" that led me to this conclusion; it had something to do with the idea that only finitely many possible values of $n?$ are less than $n!$, whereas infinitely many are greater. There's no point in reconstructing my argument, because it was totally wrong. When I wrote a program to answer the question experimentally, it became clear that nearly two-thirds of the $n?$ values are less than $n!$. (The results appear in figure 2.1.)

You can get an inkling of what's going on here by looking at a small example, say, $n = 3$, and counting the cases in which the final product is less than 3!, or 6. For $n = 3$ the only possible outputs of the random number generator are 1, 2, and 3, each of which appears with probability ⅓. The simplest event is that a 1 comes up on the first try, in which case the

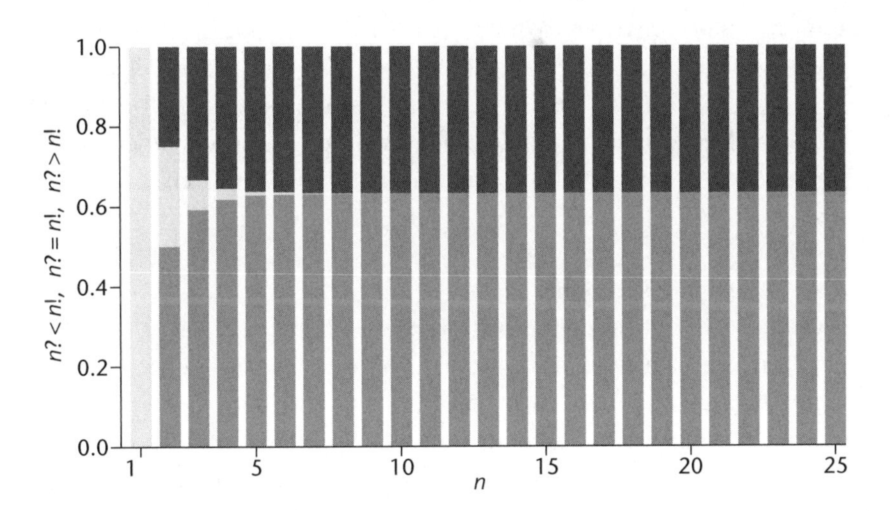

Figure 2.1 Values produced by the random factoidal function $n?$ are compared with the corresponding factorial values $n!$. For each value of n, the bars show the proportion of $n?$ values that are larger than $n!$ *(dark gray)*, equal to $n!$ *(light gray)*, and smaller than $n!$ *(medium gray)*. For large n, the proportion of $n?$ values greater than $n!$ is 0.3678, which is equal to $1/e$.

product is obviously less than 6; this happens with probability $1/3$. There are two possible sequences of just two factors, namely, a 2 followed by a 1 and a 3 followed by a 1; each of these outcomes has a probability of $(1/3)^2$, or $1/9$, and again the products are less than 6. A pencil-and-paper enumeration shows that there is only one other sequence of factors whose product is less than 6, namely, 2, 2, 1; here the probability is $1/27$. The sum of these probabilities is $1/3 + 2/9 + 1/27 = 16/27$, or 0.5926. This is in agreement with the experimental results.

As n increases, the proportion of $n?$ values less than $n!$ converges on the value 0.6322, and the proportion that are greater tends toward 0.3678. Is there anything special about these numbers, and can we explain why this particular proportion should crop up? I think so. Note that on each iteration, the probability of getting a 1 and thereby ending the factoidal process is $1/n$. That means the probability of getting anything other than a 1, and continuing the sequence of factors, is $(n-1)/n$. Then the probability of avoiding 1 twice in a row is $((n-1)/n)^2$, the probability of going on to three factors is $((n-1)/n)^3$, and so on. To have a high likelihood that $n?$ will exceed $n!$, the chain of factors has to be extended to a length of at least n. The probability of reaching this point is $((n-1)/n)^n$; as n increases, this expression converges to $1/e$, where e is Euler's number, with a numerical value of about 2.7183. The reciprocal is 0.3678—just what is observed.

If most values of $n?$ are less than $n!$, then the average of all $n?$ values should also be less than $n!$, shouldn't it? We'll see. But first another digression, to look at a closely related function that shows how averaging is *supposed* to work.

A Well-Mannered Function

If you delve into the code for the factorial function and replace the multiplication sign with a plus sign, you wind up with a procedure for calculating triangular numbers—1, 3, 6, 10, 15, 21,..., which correspond to sets of objects that can be arranged to form an equilateral triangle: • ∴ ⁂ ⁂. Whereas the factorial of 4 is $1 \times 2 \times 3 \times 4 = 24$, the corresponding triangular number is $1+2+3+4 = 10$. (The well-known shortcut for computing the nth triangular number, $n(n+1)/2$, is discussed at length in chapter 1, "Young Gauss Sums It Up." It's interesting that no such simple shortcut exists for factorials, although there are approximations.)

Having converted the $n!$ code into a program for generating triangular numbers, we can clearly make the same change in the program for $n?$. The result will be a procedure that rolls an n-sided die and keeps a running sum (rather than a product) until a 1 turns up.

Sums of random variables are much better behaved than products. With randomized triangular sums, the algorithm for calculating the arithmetic mean works flawlessly. Generate a sample of values (all for the same n), add them up, divide by the size of the sample, and you get an estimate of the mean. With a small sample, the estimate is somewhat unreliable, so that repeating the procedure is likely to produce a substantially different result.

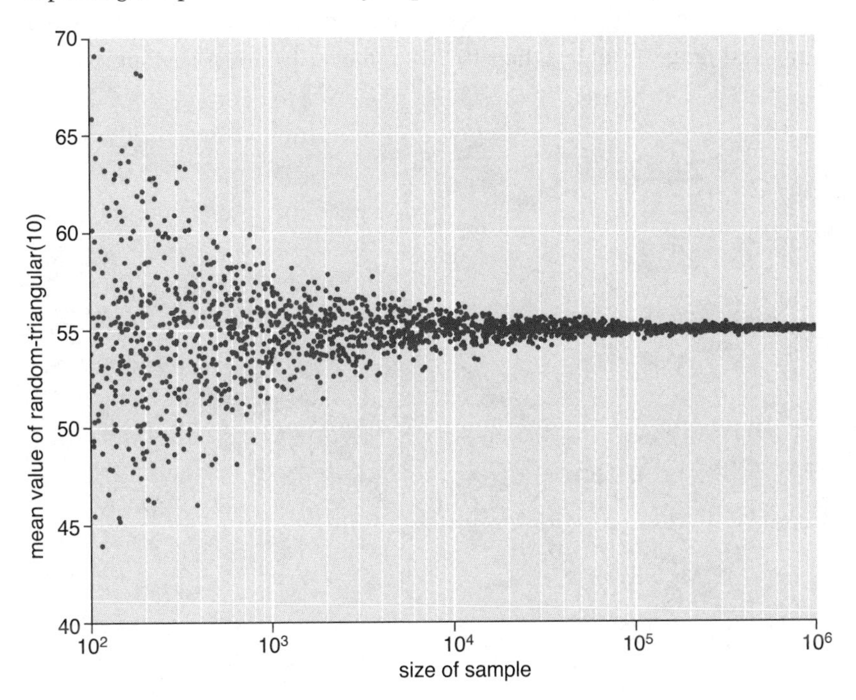

Figure 2.2 The arithmetic mean is well-defined for an additive analogue of the factoidal function—a randomized computation of triangular numbers. The algorithm selects integers randomly in the range 1 through 10, summing the numbers seen until the first 1 is drawn. The process is repeated many times, and the mean value is calculated for all the runs in the sample. For each of the 2,000 dots shown here, vertical position indicates the calculated mean, and horizontal position gives the sample size. With increasing sample size, the sample means converge to an overall mean of 55, which is the tenth triangular number.

But as the sample size increases, the estimates grow more consistent. Figure 2.2 shows the convergence of sample means toward the true mean for 2,000 samples of various sizes.

By No Means

The statistics of the factoidal process are dramatically different. When I first began playing with the *n*? function, I was curious about its average value, so I did a quick computation with a small sample—100 repetitions of 10?. The result that came back was much larger than I had expected, in the neighborhood of 10^{25}. When I repeated the computation several times, I continued to get enormous numbers, and furthermore, they were scattered over a vast range, from less than 10^{20} to well over 10^{30}. The obvious strategy

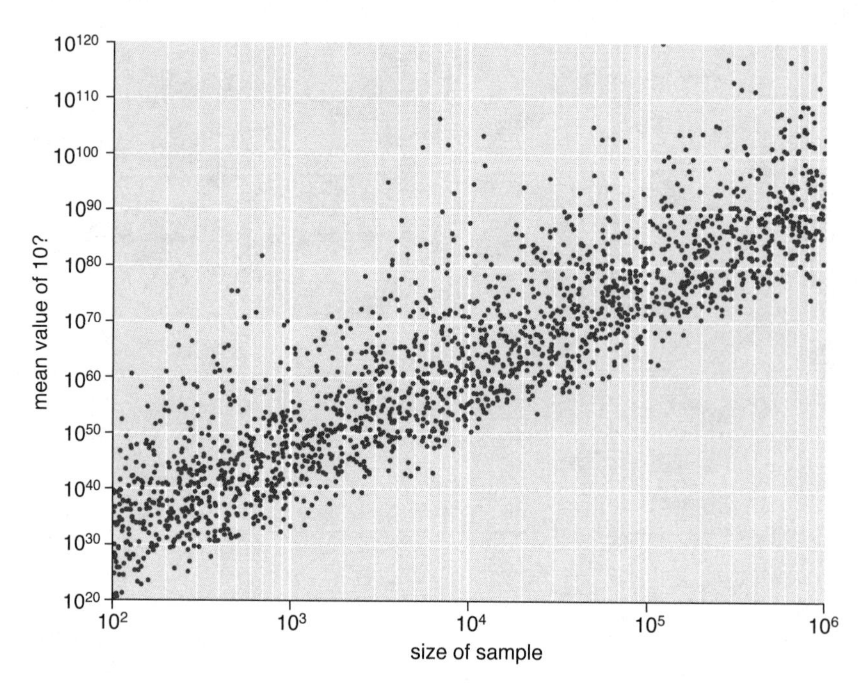

Figure 2.3 The arithmetic mean is undefined for the factoidal function. Here numbers are chosen at random from the range 1 through 10 and multiplied until a 1 is drawn. Each of the 2,000 dots is positioned according to the sample size and the calculated mean for that sample. As the sample size increases, the calculated mean does not converge to a stable value; the trend is continually rising, and so is the dispersion of the observed mean values. For samples of about 1,000 values, the mean is near 10^{50}; for samples of size 100,000, it is 30 orders of magnitude greater.

was to try a larger sample in order to smooth out the fluctuations. But when I averaged factoidals 10,000 at a time, and then 1 million at a time, the numbers got even bigger, and the variations wider.

Figure 2.3 shows what I was up against. Each dot represents a sample of runs of the *n*? program; a dot's horizontal position indicates the sample size, and its vertical position gives the arithmetic mean calculated from that sample. In all cases the value of *n* is 10. It's important to emphasize that this is *not* a graph of *n*? as a function of *n*; the value of *n* is fixed. All that changes in moving from left to right across the graph is the size of the sample over which the average is computed. There is no sign of convergence here. The trend is continuously upward; the more trials in the sample, the larger the calculated mean. The "average" value of 10? is somewhere near 10^{40} or 10^{50} if you average over 1,000 trials, but it rises to roughly 10^{90} if you go on to collect 1 million samples. (For comparison, 10! is roughly 10^6—or, more precisely 3,628,800.)

The dispersion of the dots around the trend line also shows no sign of diminishing as the sample size increases. Thus the variance or standard deviation of the data is also impossible to pin down.

Odd, isn't it? Generally, if you are conducting an experiment, or making a measurement, or taking an opinion survey, you expect that collecting more data will yield greater accuracy and consistency. Here, more data just seems to make a bad situation worse.

With a closer look at the factoidal data, it's not hard to understand what's going wrong with the computation of the mean. Although the majority of 10? values are comparatively small (less than 3,628,800), every now and then the factoidal process generates an enormous product—a rogue, a monster. The larger the sample, the greater the chance that one of these outliers will be included. And they totally dominate the averaging process. If a sample of 1,000 values happens to include one with a magnitude of 10^{100}, then even if all the rest of the data points were zero, the average would still be 10^{97}.

The arithmetic mean is not the only tool available for characterizing what statisticians call the central tendency of a data set. There is also the geometric mean. For two numbers *a* and *b*, the geometric mean is defined as the square root of $a \times b$; more generally, the geometric mean of *k* numbers is the *k*th root of their product. The geometric mean of samples taken from the factoidal process suffers from none of the problems encountered with the arithmetic mean. It converges, though somewhat slowly, to a stable

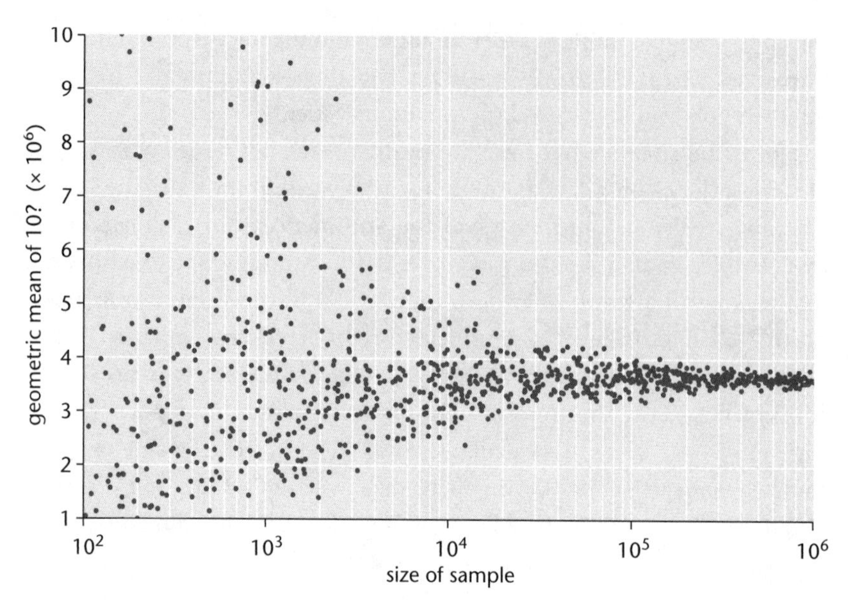

Figure 2.4 The geometric mean, unlike the more familiar arithmetic mean, is a well-behaved measure of the factoidal distribution. For a set of k numbers, the geometric mean is defined as the kth root of the product of all k numbers. For samples of 10?, the geometric mean converges slowly toward a specific, finite value as the sample size increases. The value toward which the process converges is 10!, or 3,628,800.

value (see figure 2.4). Moreover, it turns out that the geometric mean of n? is simply n!, so this is a highly informative measure. Perhaps it should not be a surprise that factoidals are better described by a statistic based on multiplication than by one based on addition.

The median of n? is also well-defined. The median is the midpoint value of a data set—the item that is greater than half the others and less than half. Because it merely counts the number of greater and lesser values, without considering their actual magnitudes, it is insensitive to the outliers that cause havoc with the arithmetic mean. For samples of 10? the median converges on a value near 27,000, notably smaller than 10! (see figure 2.5).

Still another way to tame the factoidal is to take logarithms. If you determine the logarithm of each n? value, then calculate the arithmetic mean of the logarithms, the result converges very nicely. (Note that taking the mean of the logarithms is not the same as taking the logarithm of the means.) The success of this strategy should not come as a surprise. Max Hailperin of

Gustavus Adolphus College points out that the arithmetic mean of the sum of the logarithms is the same as the logarithm of the geometric mean. Since the geometric mean is known to converge, its logarithm must do so also.

Even with other statistical methods available, it's disconcerting to face the failure of something so familiar and elementary and ingrained as the arithmetic mean. It's like stumbling into an area of mathematics where Euclid's parallel postulate no longer applies or the commutative law has been repealed. To be sure, such areas exist, and exploring them has enriched mathematics. Distributions without a mean or variance have similarly broadened the horizons of statistics. All the same, they take some getting used to.

Fat Tails

The procedure I have named the factoidal function is so simple that I'm sure someone must have noticed it before. I have not found a mention of this

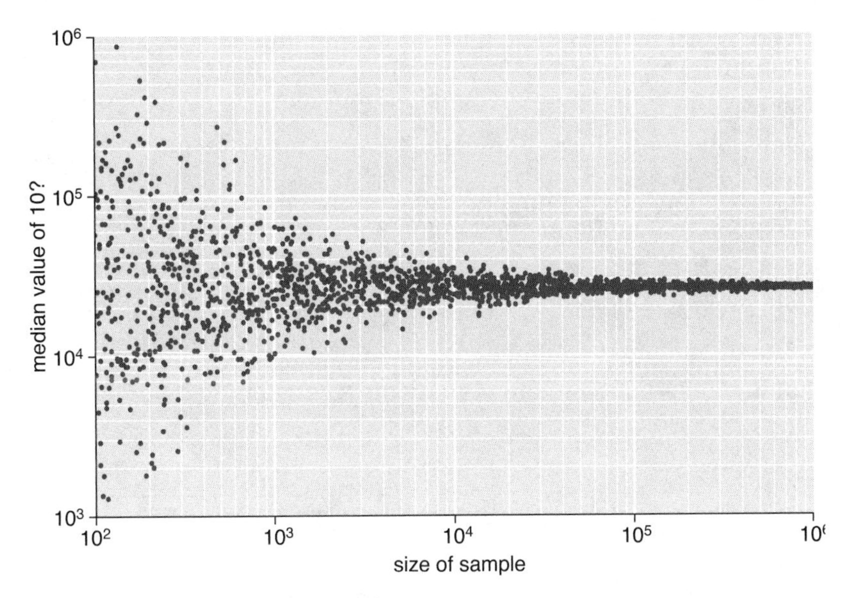

Figure 2.5 The median of the factoidal process is more informative than the arithmetic mean. The median is the midpoint value of a sample—the value chosen such that half the other values are smaller and half are larger. For data generated by 10?, the median converges to a value of about 27,000. Note that this is much smaller than 10!, which is equal to 3,628,800.

specific process, but slightly more general models involving products of random numbers do appear in the literature. (Review articles by Mark Newman of the University of Michigan and by Michael Mitzenmacher of Harvard University are particularly helpful.)

The context of these discussions is the study of heavy-tailed or fat-tailed distributions. The familiar normal distribution is *not* in this class; it is lean-tailed. The extremes of the normal probability curve, far from the peak, fall away exponentially, so that unlikely events become *really* unlikely and are never seen. Fat-tailed distributions decay more slowly, allowing room for outliers and freaks. Human height is a normally distributed variable; most people are less than two meters tall, and nobody reaches three meters. Human wealth has a fat-tailed distribution; worldwide, median net worth is about $3,600, but there are also millionaires and billionaires. (If height had the same distribution as wealth, there would be people a million meters tall.)

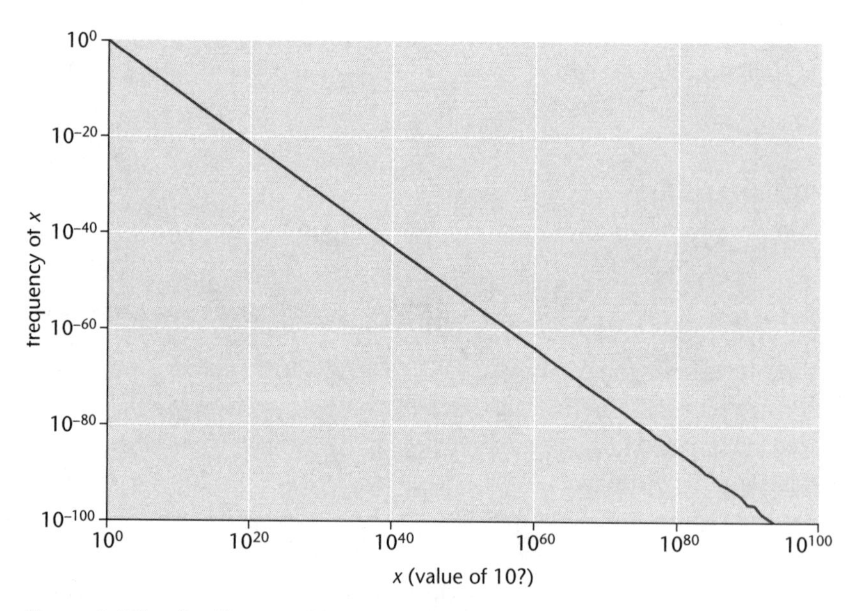

Figure 2.6 The distribution of factoidal values follows a straight line on logarithmic scales, a telltale sign of power law behavior. For each possible value, x, of the function 10?, the graph records the relative frequency at which that value of x is observed. The frequency of x is proportional to $x^{-\alpha}$, where α has a value of about 1.07. The bumpiness at the extreme tail of the distribution can be attributed to the finite sample size (10 million evaluations of the n? function).

The distribution of wealth was one of the subjects that first aroused interest in fat-tailed distributions, starting with the work of the Italian economist Vilfredo Pareto in the 1890s. Later it emerged that word frequencies in natural languages are also described by a fat-tailed distribution, usually called Zipf's law, after George Kingsley Zipf. The sizes of cities offer another example. If urban populations were normally distributed, we wouldn't have Mumbai or São Paulo. In the past decade or so, it seems like fat tails have been turning up everywhere: in the number of links to websites and citations of scientific papers, in the fluctuations of stock market prices, in the sizes of computer files.

The classic fat-tailed distribution is one where the decay of the tails is described by a power law. The probability of observing some quantity x goes as $x^{-\alpha}$, where α is a constant; the smaller the value of α, the fatter the tails. When α is less than 2, the mean of the distribution does not exist. Drawn on a graph with logarithmic scales, a power law distribution takes the form of a straight line. Another fat-tailed distribution, called the lognormal, follows a straight line over a certain range but at some point takes a sudden nosedive. The lognormal, as the name suggests, is the distribution formed by variables whose logarithms are normally distributed.

What about the factoidal function—which distribution describes the *n*? values? My first guess was a lognormal, based on a vague intuition that the logarithms of the *n*? products should indeed be normal. So much for my intuition! A log-log graph of the factoidal function (see figure 2.6) shows clear evidence of power law behavior. The graph is a straight line, with no hint of the "bended knee" to be expected in a lognormal. The calculated value of the exponent α is about 1.07, well inside the range where the mean and variance cease to exist.

With guidance from Newman and Mitzenmacher I eventually came to understand why the factoidal follows a power law. They pointed me to a paper by William J. Reed of the University of Victoria in Canada and Barry D. Hughes of the University of Melbourne in Australia. Reed and Hughes show that when a process of exponential growth is stopped at random times, the resulting distribution of values follows a power law. One of their examples is multiplication of random numbers with mean μ, stopped after a random number of terms. The factoidal function is merely a special case of this process. Later, Anthony G. Pakes of the University of Western Australia gave a thorough analysis of the factoidal process itself.

The shape of a probability distribution can have grave consequences in many areas of life. If the size and intensity of hurricanes follows a normal distribution, we can probably cope with the worst of them; if there are monster storms lurking in the tail of the distribution, the prospects are quite different. Those who make a profession of risk assessment—insurance underwriters, financial analysts—take a keen interest in these questions.

Acknowledgments

Errors and omissions in an earlier version of this essay have been corrected with the generous help of Ernst Zinner of Washington University in St. Louis; Max Hailperin of Gustavus Adolphus College in St. Peter, Minnesota; Carl Witty; and Anthony G. Pakes of the University of Western Australia.

3

How to Avoid Yourself

Every Sunday morning you go for a walk in the city, heading nowhere in particular, with just one rule to your rambling: you never retrace your steps or cross your own path. If you have already walked along a certain block or passed through an intersection, you refuse to set foot there again.

This recipe for tracing a loopless path through a grid of city streets leads into some surprising back alleys of mathematics—not to mention byways of physics, chemistry, computer science, and biology. Avoiding yourself, it turns out, is a hard problem. The exact analysis of self-avoiding walks has stumped mathematicians for half a century. Even counting the walks is a challenge. If you want to take a long self-avoiding walk—say, 100 city blocks—nobody knows how many ways you can do it.

My own initiation into the trials of self-avoidance came when I began experimenting with a simple model of the folding of protein molecules. Protein folding is close to the historical roots of the self-avoiding walk, which was first conceived as a tool for understanding the geometry of long-chain polymer molecules. A polymer writhing and wriggling in solution like a strand of spaghetti forms a random tangle—random, that is, except that no two atoms can occupy the same position at the same time. This "excluded volume effect" in the polymer is modeled by the walk's insistence on avoiding itself.

Self-avoiding walks have also found applications elsewhere in the sciences, such as the physics of magnetic materials. Some foraging animals leave trails that resemble self-avoiding walks, presumably because there's no point in grazing where the grass has already been eaten. And the walks are of interest as purely mathematical objects. Many questions about them have resisted rigorous analysis, and so the best answers known so far come from computer-intensive experiments.

Most of the walks I describe here take place on a two-dimensional square lattice, which is a grid of city streets reduced to its mathematical essence. The lattice consists of all points on the plane that have integer x and y coordinates. Walks begin at the origin, the point with coordinates $x = 0$ and $y = 0$. A single step always moves from the current lattice site to one of the four nearest-neighbor sites. By convention the length of a walk, n, is defined as the number of steps, and so the number of lattice sites visited is $n+1$.

I Wonder as I Wander

In trying to understand the self-avoiding walk, a good place to begin is with a walk that doesn't bother to avoid itself but lurches over the landscape entirely at random. At each step of such a walk you choose one of the four neighboring lattice sites with equal probability, and move there. If you repeat this process a few hundred times, drawing a line behind you as you go, the result is a scribble with a random but nonetheless distinctive and recognizable geometry.

How many different random paths can be traced on a square lattice? From the point of origin there are just four walks consisting of a single step, namely, those going one unit north, east, south, or west. On the second step each of these walks can be continued in any of the same four directions, and so there are 16 two-step walks. For every further step the number of walks is again multiplied by 4, so that the number of n-step walks is 4^n. The sequence begins 4, 16, 64, 256, 1,024 . . . .

An interesting question about random walks is whether the walker ever returns to the starting point. Almost 100 years ago George Pólya showed that the answer depends on the nature of the lattice, and specifically on how many spatial dimensions it fills. In one or two dimensions a random walker is certain to come back home if the walk continues long enough; the

Figure 3.1 Three kinds of walks explore a rectangular grid of streets. A random walker can choose any of four directions at each intersection; a nonreversing walker cannot make a U-turn and so has three choices at every site after the first; a self-avoiding walker cannot return to any visited site. Each walk shown here consists of 1,000 steps, starting at the open circle and ending at the filled one. The self-avoiding walk is shown at reduced scale because it covers a larger territory.

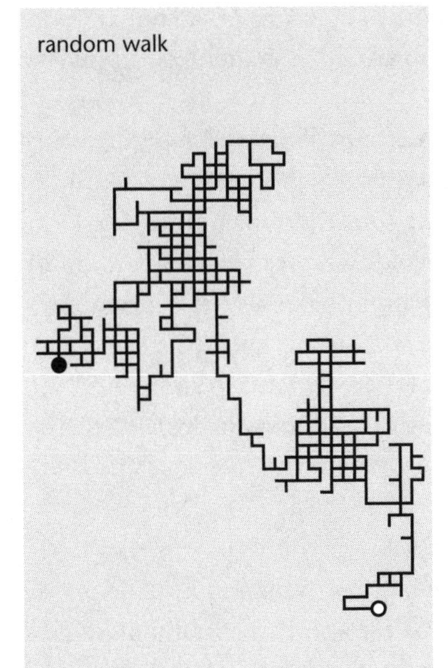
random walk

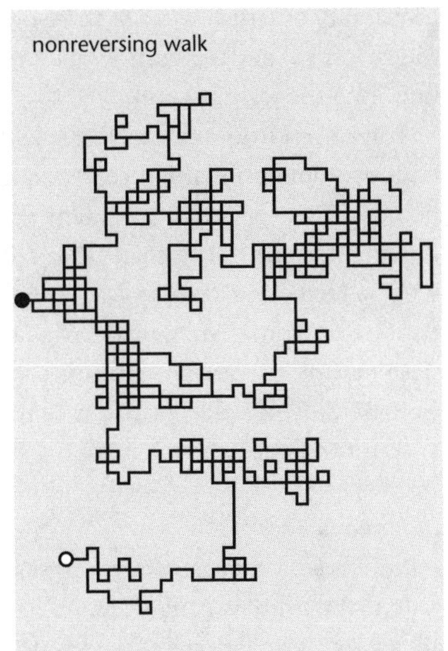
nonreversing walk

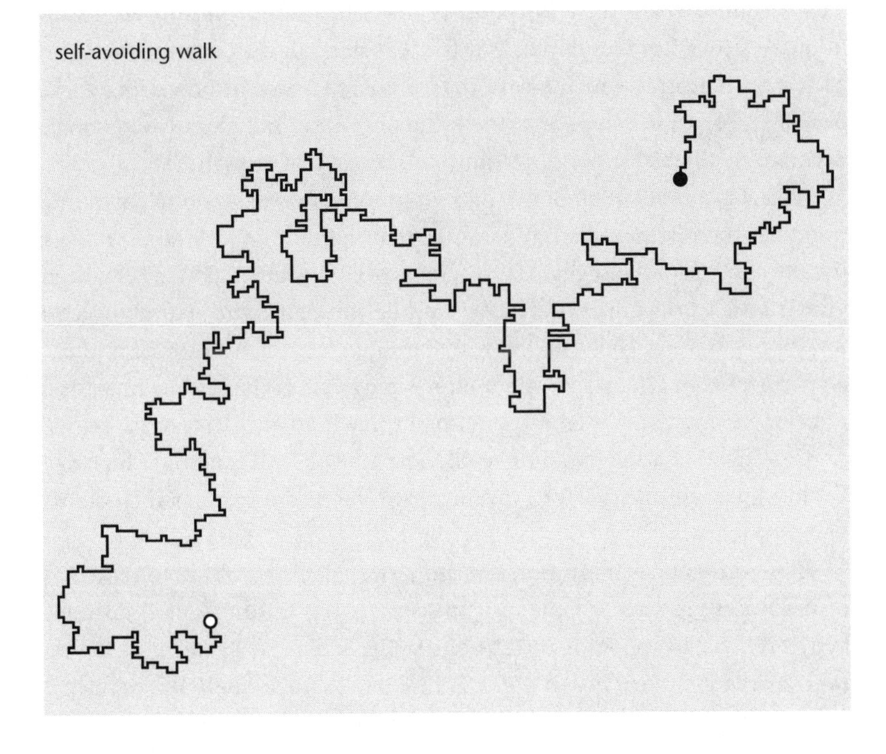
self-avoiding walk

probability of return is 1. But three or more dimensions offer enough room to get lost in, and a return to the origin cannot be guaranteed no matter how long the walk goes on.

Pólya's result immediately tells us something about self-avoiding walks on a two-dimensional lattice. If a random walk's probability of returning home is 1, the probability of *not* revisiting the origin must be 0. And since the origin is one of the places that self-avoiding walks avoid, an arbitrarily long self-avoiding walk must be highly improbable—so rare and exceptional that you have almost no chance of finding one. This scarcity of specimens is one reason self-avoiding walks are so hard to study. And yet, paradoxically, another reason is that they're so numerous it's a challenge to count them all.

Don't Look Back

An intermediate stage between a purely random walk and a self-avoiding walk is the nonreversing walk. As a recipe for an urban perambulation, it allows you to go left, right, or forward at each intersection, but not to make a U-turn and go back the way you just came. Thus a nonreversing walker on a square lattice has four choices for the first step but only three choices for each step thereafter, which means the number of n-step nonreversing walks is $4 \times 3^{n-1}$. For large values of n, the influence of the initial four-way choice becomes negligible, so we can simply call the rate of growth 3^n.

Typical examples of random walks, nonreversing walks, and self-avoiding walks can be distinguished at a glance (see figure 3.1). The random walk usually consists of dense regions, where most of the lattice points have been visited at least once, connected by tendrils through more sparsely settled territory. A trace of the walk looks something like a map of towns and cities connected by highways. The nonreversing walk is similar but suggests a more open landscape—perhaps suburban sprawl rather than the city center. And the trace of a self-avoiding walk looks not like cities along a highway but like the highway itself, or a meandering river. There are no branch points or closed loops.

These differences in appearance are reflected in quantitative measures of the walks' geometry. One such measure is the mean squared distance between the starting point and the end point of a walk. To calculate it, you need a large sample of n-step walks. For each walk, measure the straight-line distance from the origin to the end point, square this number, and then take

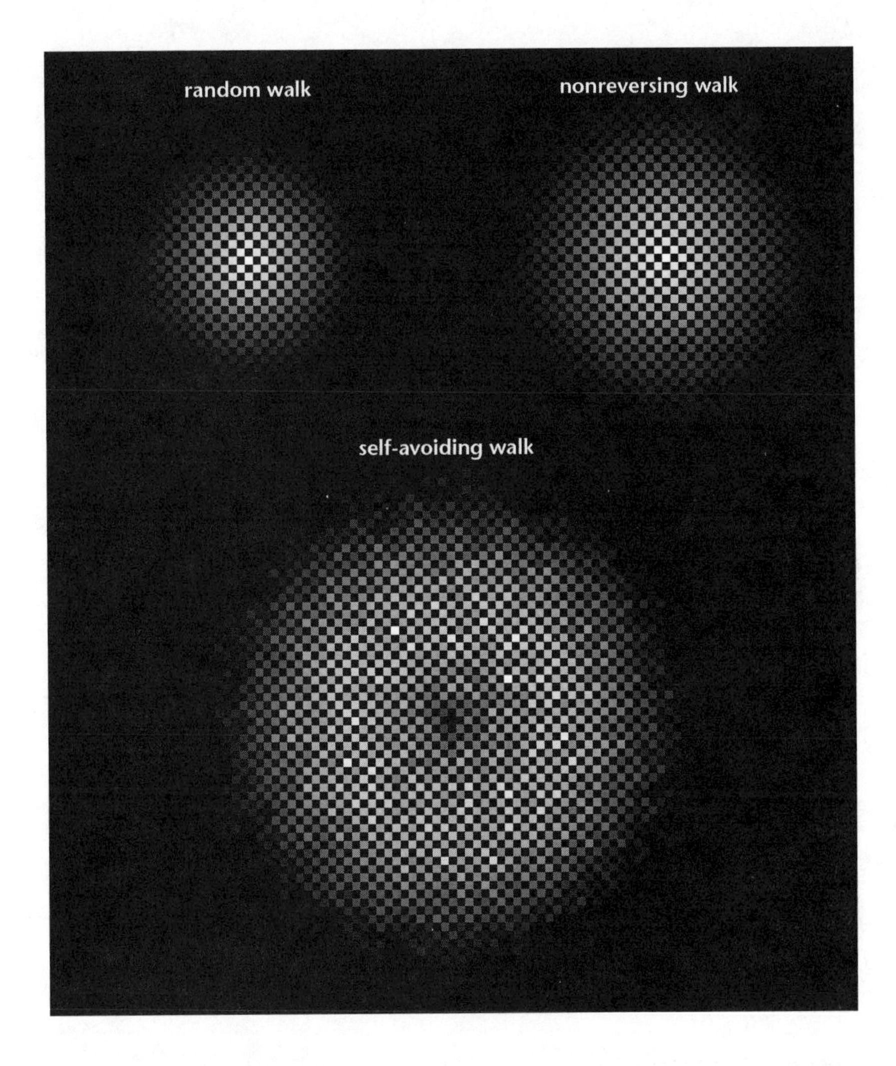

Figure 3.2 The distance covered by walks on a square grid, and hence the distribution of the walks' end points, depends on the rule governing the walk. Each of the three images records the end points of 10,000 walks of 50 steps each. The walks begin at the origin ($x = 0$, $y = 0$) at the center of each disk. The number of walks ending at any lattice site is encoded in the brightness of the corresponding square. Random walks are the most compact, nonreversing walks a little less so. In both cases the likeliest place for a walk to end is at or near the origin. Self-avoiding walks extend over a wider territory; moveover, they tend to avoid returning to the neighborhood of the origin, giving a doughnut-like distribution of end points. The checkerboard motif common to all three images arises because a 50-step walk must end at a site whose $x+y$ distance from the origin is an even number.

the average over all the walks. For random walks the average of the squared displacement is n. (This does *not* mean that the average distance between the starting point and the ending point is $\sqrt{n}$, because the average of the squared distance is not the same as the square of the average distance.) For nonreversing walks the corresponding mean squared displacement is $2n$. Self-avoiding walks are qualitatively different. The mean squared displacement grows as a nonlinear function of n, which appears to be $n^{3/2}$.

Roughly speaking, the mean squared displacement measures the size of the territory covered by a walk. Consider a large sample of 100-step walks. If the walks are purely random, the mean squared displacement will be close to 100, and a typical walk will be bounded by an area of 100 square blocks. For nonreversing walks the area will be about 200 square blocks. In the case of self-avoiding walks, the expression $100^{3/2}$ works out to 1,000. Thus the effect of avoiding your own path is to stretch out your walk over a much larger territory (see figure 3.2).

There is another important difference between random walks and self-avoiding walks. Every random walk or nonreversing walk can go on forever; you can always take one more step. But a self-avoiding walk can stumble into a blind alley, getting trapped at a lattice site where none of the neighbors are unvisited. In other words, sometimes you can't avoid yourself no matter how hard you try. On any given step the probability of getting boxed in is small—a little less than 1 percent—but if you extend a walk indefinitely, it is certain to wander into a dead end eventually. This is another way of saying that self-avoiding walks are rare and special; they have to beat the odds just to survive.

Counting Your Steps

How many distinct n-step self-avoiding walks can you take on a square lattice? There is no known exact formula, analogous to the expression 4^n for random walks or $4 \times 3^{n-1}$ for nonreversing walks. The closest we can come to a formula is to set upper and lower bounds. The number of self-avoiding walks has to be less than 3^n because that is the number of nonreversing walks, which include the self-avoiding walks as a subset. Similarly, it's easy to construct subsets of the self-avoiding walks whose numbers grow as 2^n; an example is the family of walks that move only north or east at each step.

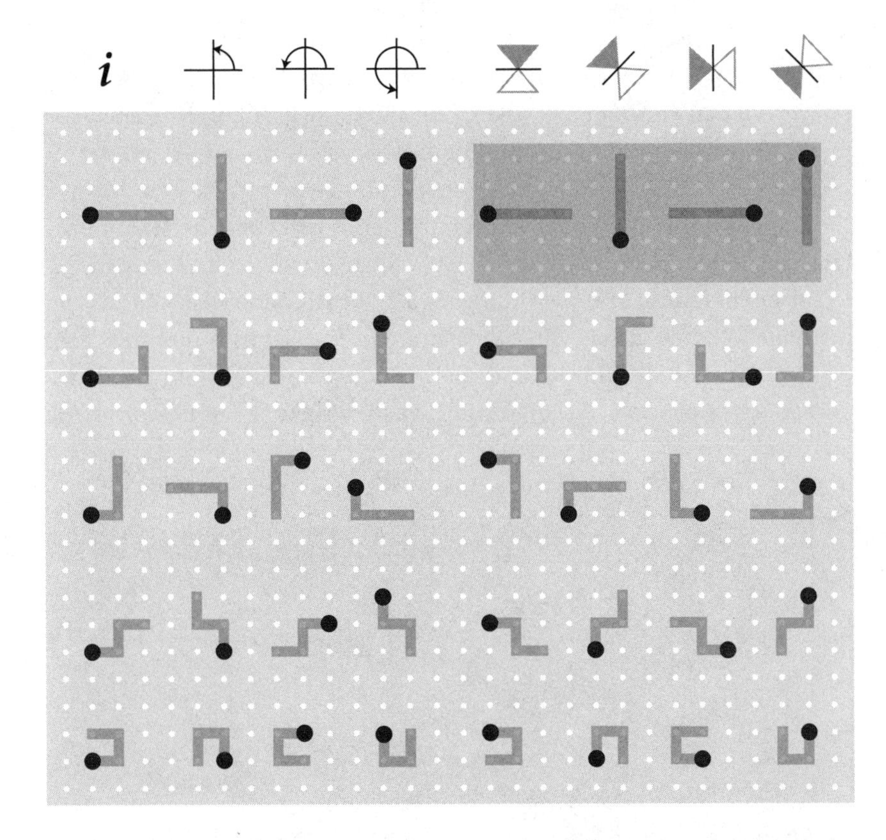

Figure 3.3 Symmetries reduce the number of unique walks by a factor of about eight. Five unique three-step walks are shown in the leftmost column (which is marked *i* for identity). In the other seven columns the walks are rotated by 90, 180, and 270 degrees, and reflected across four mirror axes (vertical, horizontal, and two diagonals). There are 40 walks altogether, but for straight walks (first row) reflection has the same effect as rotation; excluding the duplicates in the dark gray box leaves 36 distinguishable walks.

Thus the number of n-step self-avoiding walks must lie somewhere between 2^n and 3^n. Empirical evidence from computer enumerations indicates that the growth rate is about 2.638^n.

If you want an exact tally of the n-step self-avoiding walks, the only known way to get it is to actually count them. You can make a start with a pencil and a sheet of squared graph paper. There are four one-step walks, proceeding north, east, south, and west from the origin. Each of these walks can be extended in any of three directions (not four, because the walker can't do an about-face and go back the way it came). Thus there are $4 \times 3 = 12$ two-step walks. Continuing in this way, you can extend the 12 two-step walks to generate 36 three-step walks.

At about this point, you may begin to notice that the process is annoyingly repetitious. Essentially the same walk appears over and over in different orientations. As shown in figure 3.3, a path on a square lattice can be rotated to face in four directions and reflected across four mirror axes (vertical, horizontal, and two diagonals). Should the results of these transformations be considered eight distinct walks or eight variations on a single

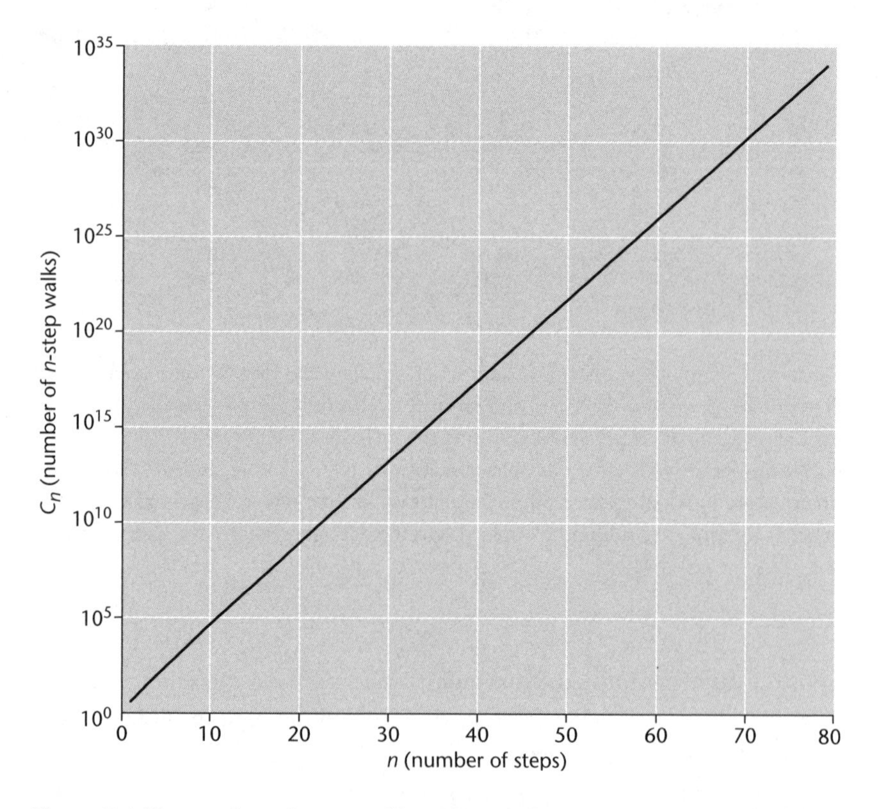

Figure 3.4 The number of *n*-step self-avoiding walks grows exponentially with *n*. There are just four one-step walks but more than 10^{34} 79-step walks. In this graph the vertical scale is logarithmic, which means the exponential curve appears as a straight line.

walk? The answer depends on what you want to do with the walks, but if you're merely counting them, it's foolish to count by ones when you can count by eights. To eliminate the fourfold rotational symmetry, you can generate only those walks that start with a step in some particular direction, say, east. The four mirror symmetries disappear if you consider only walks whose first turn after the initial step is in one specified direction, say, north. In this way the number of walks to be counted is reduced to approximately one-eighth the total number. Why approximately? Because of one small complication: a straight walk makes no turns away from its initial direction and is therefore left unchanged by mirror reflection. Hence there are only four distinguishable straight walks instead of eight, and the total number of walks is reduced by four.

Up to this point, the prohibition on self-intersection hasn't actually entered the calculation, except by forbidding U-turns. On a square grid, a one-, two-, or three-step walk simply can't run into itself. But beginning with four steps, the walker can cross its own trail, and so the counting algorithm must be careful to exclude all self-intersecting pathways. It turns out there are 14 unique four-step walks (excluding rotations and mirror reflections), but one of them is self-intersecting, namely, the square walk that goes east, north, west, and south, returning to the origin. The 13 remaining four-step walks yield $13 \times 8 - 4 = 100$ self-avoiding walks when the symmetries are taken into account.

In 1959 Michael E. Fisher and M. F. Sykes enumerated all the two-dimensional square-lattice walks with up to 16 steps, which was a remarkable accomplishment with their paper-and-pencil methods. (At $n = 16$ there are 17,245,332 walks.)

If you want to duplicate this feat or go beyond it, a computer is highly recommended. The simplest walk-counting program generates all n-step nonreversing walks (excluding rotational and mirror symmetries), then discards those that have self-intersections. How does the program detect self-intersections? In order to avoid your own footsteps, you must somehow remember where you've been. In one strategy, with each step of the walk you add the coordinates of the site you have just reached to a list of x,y pairs. Then the walk is self-avoiding if there are no duplicated pairs in the list. Another approach is to set up a two-dimensional array representing the region of the lattice accessible in n steps. As each new site is reached, you leave a marker (call it a breadcrumb) at the corresponding site in the array.

Then any self-intersection will be apparent immediately; when you go to leave a breadcrumb, you'll find there's one already there. The breadcrumb scheme is faster because with each step of the walk, you have to check only one element of the array for a conflict—the element indexed by the coordinates of the new lattice site.

Still another family of algorithms begins with the observation that any self-avoiding walk of $2n$ steps can be viewed as the concatenation of two n-step walks. Suppose you generate and store all 44,100 walks of 10 steps each. You can then combine the walks in pairs, leaving the first walk (call it A) in its original position, and translating the second walk (B) across the lattice so that B's starting point coincides with A's end point. Each B must be tried in each of three orientations. The advantage of this strategy is that you already know that both A and B are self-avoiding within their own lengths; you only need to check for collisions between A and B. The procedure is still quite arduous. In concatenating two 10-step walks, there are $3 \times 44{,}100^2$ combinations to be checked, or almost 6 billion. After the self-intersecting walks are removed, 897,697,164 self-avoiding 20-step walks remain. This method of building walks by doubling is known as dimerization, after the process in polymer chemistry where monomers pair up to form dimers.

With algorithms like these you can easily count all the 10- and 20-step self-avoiding walks. Maybe, given patience and skill and a fast computer, you could even generate all the 30-step walks. (There are 16,741,957,935,348 of them.) But to go much further you would need more sophisticated tools.

The Higher Arithmetic

Over the past 50 years, most of the industrial-scale work on the counting of self-avoiding walks has been done by a group at the University of Melbourne in Australia, whose members have included Anthony J. Guttmann, A. R. Conway, Ian G. Enting, Iwan Jensen, and others, along with collaborators elsewhere. In 1987 the Guttmann group reached $n = 27$, then soon after raised the bar to 29. Others reported results for 30 and 34 steps, and Guttmann's group went on to 39. In 1996 Conway and Guttmann enumerated all the self-avoiding walks through $n = 51$; there are 14,059,415,980,606,050,644,844 of them.

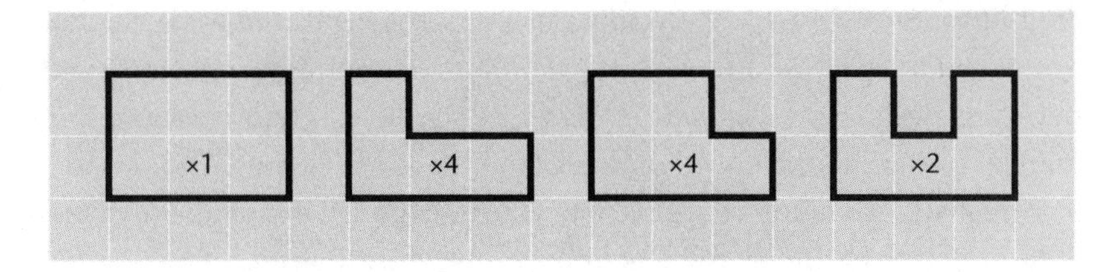

Figure 3.5 Four distinct self-avoiding polygons fit inside a 2 × 3 rectangle and touch all four sides of the bounding box. Allowing for rotations and reflections, there are 11 such polygons overall. The first three contribute to the count of self-avoiding polygons of perimeter 10. The last polygon has perimeter 12.

In 1998, when I first wrote on this subject, I remarked that "exhaustive enumerations of self-avoiding walks seem unlikely to advance much beyond the current limit of $n = 51$." My pessimism was unfounded. Six years later Jensen completed counts of all self-avoiding walks through $n = 71$, and he has since pushed on to $n = 79$. The number of 79-step walks is a staggering 10,194,710,293,557,466,193,787,900,071,923,676, or about 10^{34} (see figure 3.4)

In recent years these gargantuan enumerations have been done on machines ranked as supercomputers or on clusters of workstations with hundreds or thousands of processors and many gigabytes of memory. And yet fancy hardware is not the primary key to these accomplishments. Algorithmic ingenuity has made a far larger contribution.

The most direct and obvious methods of enumerating walks organize the task hierarchically. You begin with all possible one-step walks, then consider all possible ways of extending those walks to two steps, and so on. The algorithms behind the record-setting counts break the process down differently. They examine finite regions of the lattice—in the case of the square lattice the regions are always rectangles—and tally how many different walks can be embedded in each of those regions. Summing up contributions from all regions that could bound a walk of a given length yields the grand total.

The finite-lattice family of algorithms was devised in the early 1980s by Enting, who was then at Northeastern University. It was originally applied to self-avoiding *polygons*, which are self-avoiding walks that return to the origin to form a closed loop. In one version of the algorithm, the polygon is required to touch all four edges of the bounding rectangle. Then there's never any need to consider rectangles whose perimeter is greater than n, the number of steps in the polygons being counted.

For small rectangles, counting the edge-to-edge self-avoiding polygons is easy. Figure 3.5 shows the four unique polygons that fill a two-step by three-step box. When various rotations and reflections are included, there are 11 of these closed paths. Larger rectangles demand a systematic scheme for tabulating the legal paths. Enting devised such a scheme, based on certain constraints that must be observed throughout the construction of a polygon. First, every vertex of the lattice must be touched either by zero edges or by exactly two edges. This is a strictly local condition; you can

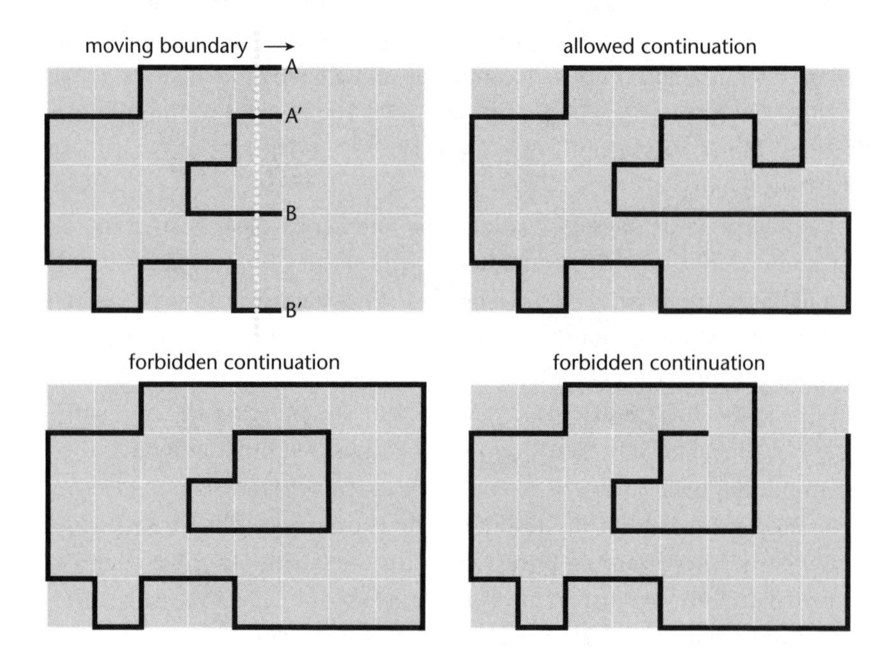

Figure 3.6 A finite-lattice algorithm constructs self-avoiding polygons by enforcing local and global constraints on the developing pathway. In the panel at upper left, a moving boundary *(dotted line)* separates partially complete pathways from the remaining blank canvas. The path is continued by building onto four open segments, labeled *A*, *A′*, *B*, *B′*. Only the continuation where *A* connects to *A′* and *B* to *B′* leads to a valid self-avoiding polygon.

tell whether the rule is met by looking at a single vertex and its immediate neighborhood. The second requirement is that all the edges link up to form a single, noncrossing loop. This constraint is more global, and yet it turns out that local configurations convey useful information about the global connectivity of the path.

In the finite-lattice algorithm a moving boundary line scans across the lattice from left to right, separating completed parts of the pathway on the left from the undefined region on the right. At any intermediate stage the local and global constraints limit the choices that will lead to a valid, complete polygon (see figure 3.6.). The continuation must fulfill the local requirement that every vertex touch zero or two edges, and the completed path must form a single closed loop. By ruling out impossible configurations, the algorithm avoids wasting time on continuations that cannot contribute to the final count of polygons.

In the 1990s Enting's algorithm for polygons was extended to open-ended self-avoiding walks by Conway, Enting, and Guttmann. Walks are more difficult than polygons for two reasons. First, walks spread out more than polygons do, and so the algorithm must survey larger rectangles. Second, the simple rule that all vertices must meet either zero or two edges is no longer valid; two special vertices, at the end points of the walk, have just one adjacent edge, which complicates the bookkeeping necessary to enforce the constraints.

Algorithms in the finite-lattice family are memory-hungry and difficult to implement correctly, according to their authors. (I haven't tried it myself.) But there's a major payoff in efficiency. Programs based on direct enumerations have a running time that increases exponentially in proportion to 2.638^n. The finite-lattice method still shows exponential growth but at a much lower rate. The number of operations is proportional to $3^{n/4}$, which is equivalent to about 1.334^n. For $n = 79$, that's the difference between 10^{34} and 10^{10}.

Going Deeper

Why invest so much hard work in counting self-avoiding walks? It's not just for the idle joy of calculating ridiculously large numbers. Exact enumeration offers the best hope of understanding the large-scale behavior of self-avoiding walks, of being able to mathematically describe and summarize how these wriggling paths arrange themselves in space without ever touching the same point twice.

I have mentioned that there is no exact formula for the number of n-step walks, but there is a formula that is thought to become exact in the limit of infinite n. This asymptotic formula looks like this:

$$C_n = A\mu^n n^g.$$

Here C_n is the count of n-step walks. The coefficient A and the exponent g are universal constants, so named because they are the same for all lattice geometries in the same spatial dimension. In other words, they have the same values for any plane lattice—squares, triangles, or hexagons—but the values are different for three-dimensional lattices. The number μ is called the connective constant, or simply the growth constant, of the self-avoiding walk, and its value *does* vary depending on the details of the lattice shape. In two dimensions μ has one value for a square lattice, another for a triangular lattice, and so on.

On the plane, the universal exponent g is believed to have the exact rational value 11/32. This fact has not been proved, but it is supported by an abundance of evidence and plausible theoretical arguments. The coefficient A has no such exact value; it has to be determined from measurements of the geometric properties of self-avoiding walks, such as the mean squared end-to-end distance. The estimated value of A is about 1.77.

The dominant contribution to the number of walks is the exponential factor μ^n; for any sufficiently large value of n, μ^n exceeds the power law factor n^g. The connective constant μ represents the average number of choices available to a self-avoiding walk—the number of ways of continuing the walk without violating the self-avoidance condition. In the case of the square lattice, it is a number we have already encountered. As mentioned, there are simple arguments that require μ on the square lattice to lie somewhere between 2 and 3, and the value found empirically is about 2.638. As a matter of fact, the main reason for performing all those heroic enumerations of self-avoiding walks is to gather data leading to more precise estimates of μ (see figure 3.7).

The foregoing formula for C_n seems to offer an easy way to calculate μ. Take any known value of C_n, plug in the best estimates of A and g, then turn the equation inside out to solve for μ. This procedure would work if we knew C_n for some really large values of n—those approaching the neighborhood of infinity. But 79 is not particularly close to infinity, and so more elaborate statistical methods of estimation are needed. Careful examination of how estimates of μ change as a function of n have produced values accurate to 12 significant figures.

Early in the study of self-avoiding walks, there was speculation that μ on the square lattice might have some special distinguished value. One author suggested it could be $1+\sqrt{2}$, equal to about 2.414. Another proposed that μ

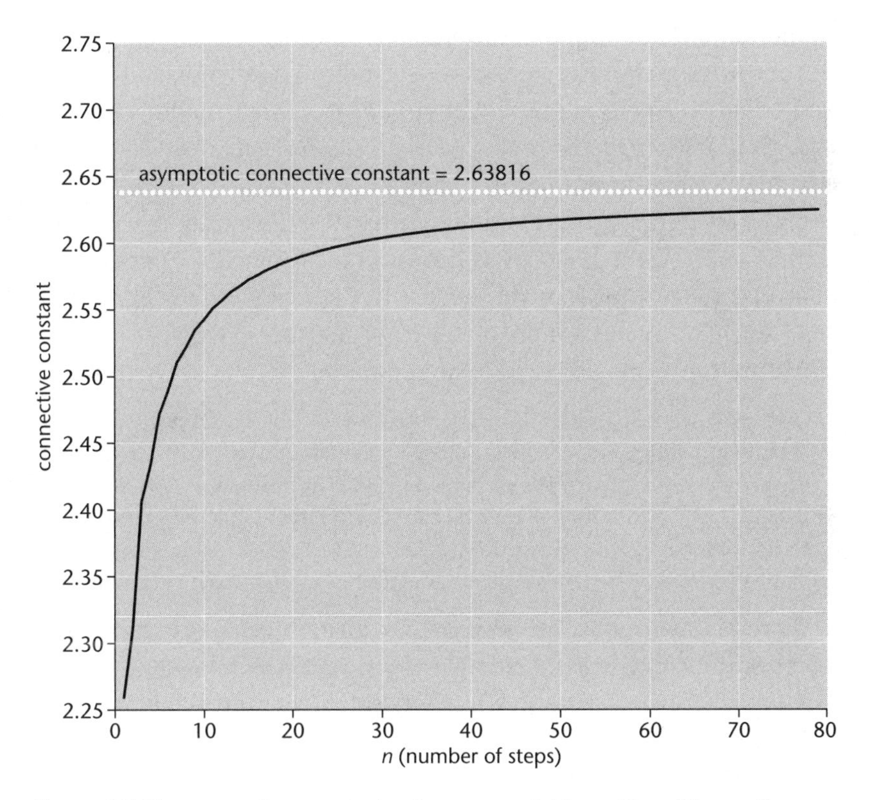

Figure 3.7 The connective constant μ for a square-lattice self-avoiding walk represents the average number of choices available at each lattice site, and determines the rate of growth in the number of walks as the number of steps n increases. Estimates of μ based on enumerations of walks with up to 79 steps *(black curve)* are slowly approaching the asymptotic value of 2.63816 *(dotted white line)* predicted to hold for very large n.

is equal to e, or Euler's number, the base of the natural logarithms, with a value of roughly 2.718. More careful measurements based on enumerations of walks soon ruled out both these values (the first being too low and the second too high).

And yet it is not such a kooky idea that μ might be a number with some simply expressed form. On the hexagonal lattice (also known as the honeycomb lattice, since that's what it looks like), μ takes on the exact value $\sqrt{2+\sqrt{2}}$, equal to about 1.848. This curious identity was first conjectured by Bernard Nienhuis in 1982 and was finally proved in 2012 by Hugo Duminil-Copin and Stanislav Smirnov.

The honeycomb connective constant $\sqrt{2+\sqrt{2}}$ is not a rational number, but it is an algebraic number—a solution to a polynomial equation with integer coefficients (namely, $t^4-4t^2+2=0$). Inspired by this observation, Guttmann mounted a search for an algebraic number that would match the observed value of the square-lattice connective constant. In the 1980s he conjectured that μ for the square lattice is a root of the equation $13t^4-7t^2-581$. At the time, μ was known to six significant figures, with a value of 2.63816. This was consistent with the root of Guttmann's equation, which to 20 significant figures is 2.6381585303417408684.

In 2016 Guttmann and two colleagues wrote:

> Over the intervening years, indeed decades, as the estimate of the value of μ became increasingly more precise, this polynomial root continued to satisfy the current best estimate. For example, in 2001, Guttmann and Conway . . . quoted $\mu = 2.638158534(4)$ as the best current estimate. . . . Eleven years later, Clisby and Jensen . . . estimated $\mu = 2.63815853035(2)$. So the original quadratic mnemonic based on a 6-digit estimate of μ is seen to hold for 12 digits.

However, after a reign of 30 years, Guttmann's conjecture has finally been overthrown. The latest estimate of μ is 2.63815853032790(3), which differs from the predicted value in the twelfth digit. Apparently it was all just a coincidence.

Random Self-Avoidance

I am not going to repeat my foolhardy claim that exhaustive enumeration of self-avoiding walks has reached its limit. Some new computing technology or an algorithmic innovation may allow us to push on well beyond $n = 79$. Until that happens, however, our best hope for exploring longer self-avoiding walks is random sampling—giving up the aim of generating every n-step walk and looking at just a representative subset. Even this process is computationally intensive.

In selecting a random sample, it's important to ensure that every possible n-step self-avoiding walk has an equal chance of being included. This can be tricky. You can build an n-step walk one step at a time by choosing directions at random, but what do you do if the walk collides with itself before reaching n steps? The temptation is simply to back up one step and try another direction, but that practice leads to a biased sample of walks. To ensure a fair sample, you have to abandon a failed walk entirely and start over.

Algorithms based on this rejection protocol readily produce large samples of 60- or 70-step walks, or smaller numbers of 100-step walks. As the walks get longer, however, the proportion of candidates that pass the self-avoidance test declines sharply. At $n = 100$ you need to propose more than 50,000 walks, on average, for every one that turns out to be self-avoiding. At $n = 200$ the acceptable walks are rarer than one in a billion.

Other algorithms extend the range of exploration into the thousands of steps. In 1969 Zeev Alexandrowicz of the Weizmann Institute of Science suggested the method of dimerization, already described in the context of exact enumerations. Dimerization works because it's much easier to create two 50-step walks than a single 100-step walk. You build the two shorter walks and string them together end-to-end. Of course, the two half-walks may collide, in which case you have to start over, but failure is much less likely than in the step-by-step technique. The procedure can be invoked recursively to build the 50-step walks from 25-step components, and so on. What's particularly sweet about the dimerization algorithm is that it lends itself to a very simple and transparent implementation; I found it easier to get right than the less efficient step-by-step methods. The 1,000-step self-avoiding walk shown in figure 3.1 was generated by dimerization.

Another technique, called the pivot algorithm, also goes back to 1969; it was first described by Moti Lal of the Unilever Research Laboratory and was later refined and extended by Neal Madras of York University and Alan D. Sokal of New York University. The pivot algorithm is quite different from all the others described here. It does not actually generate a self-avoiding walk but instead takes one walk and transforms it into another. The idea is to randomly choose a pivot point somewhere along the walk, and then rotate or reflect or reverse the segment on one side of the pivot. If the result is a self-avoiding path, the new walk becomes the starting point for the next pivoting operation; otherwise you keep the original walk and choose a new pivot. Successive walks in the sequence are highly correlated, but repeating the transformation many times wipes out all memory of former configurations.

Random sampling can yield an estimate of the number of n-step walks. The rejection rate offers one way of performing the calculation. If you generate lots of n-step random walks and find that one in 1 million is self-avoiding, you can divide the total number of walks (4^n) by 1 million to estimate the size of the self-avoiding subset. In principle, you might also try the mark-and-recapture method popular in wildlife studies. The biologist catches 100 fish, tags them, and returns them to the pond. Then she catches another 100, notes that 10 of them were tagged in the first round, and concludes that the total population is 1,000. The same scheme could be applied to self-avoiding walks, but the rarity of long walks puts the practicality of the method in doubt.

The longer walks produced by random sampling could someday aid in calculating a more precise value of the growth constant μ. They bring us closer to the asymptotic region, where the formula for the number of walks becomes exact as n goes to infinity. But sampling also introduces statistical uncertainties, caused by the finite sample size. So far, those uncertainties are large enough that complete enumerations, even of much shorter walks, still give better results.

Getting Rigorous

Computational studies of self-avoiding walks have produced a rich harvest of empirical results. Theorems have been harder to come by. For example, studies of the mean squared end-to-end displacement, based on both complete enumerations and on random samples, strongly support the hypothesis that the displacement grows as $n^{3/2}$. Indeed, everyone "knows" that this result is correct and exact. But so far no one has proved it; no one has even proved that the exponent must be greater than 1 or less than 2.

Rigorous results on the counting of self-avoiding walks are also scarce. The asymptotic formula for the number of walks, $A\mu^n n^g$, seems to successfully describe what happens as n goes to infinity, and by appending various correction terms it can be made to work for finite n. Still, the growth law has not been explained from first principles, and the exact value of μ remains a mystery. For several years, it wasn't even certain that the number of walks invariably increases as n gets larger; because walks can become trapped, it seemed possible that there might be some range of values where there are fewer $(n+1)$-step walks than n-step walks. In 1990, however, George L. O'Brien proved that the series increases monotonically.

Even if the asymptotic growth law is correct, it is only an approximation—perhaps good enough for chemists and physicists but not wholly satisfying to mathematicians. Ideally, one would like a formula for calculating the exact number of walks for any value of n, without all the laborious counting. Is that too much to ask? Most likely it is. Conway and Guttmann have given compelling arguments (though not quite a proof) that no simple analytic function predicts the exact number of self-avoiding walks.

Perhaps the absence of such a function tells us something important about the nature of self-avoiding walks. The number of n-step walks is perfectly definite and knowable; there is nothing random or uncertain about the number of ways to arrange a nonintersecting path on a lattice. So why can't we calculate it? I don't know the answer, but I would point out that there are many objects in mathematics that exhibit the same curious mixture of determinism and unpredictability. The prime example is the prime numbers. Again there is nothing uncertain or statistical about what makes a number prime, but if there is any pattern in the distribution of the primes, it remains totally inscrutable. As with the self-avoiding walks, there are good approximations for the number of primes in a range of numbers, but no one has found an exact formula that reliably points to every prime. This stubborn resistance to total analysis is part of what makes the primes interesting. Perhaps self-avoiding walks belong in the same category of perpetually tantalizing mathematical structures.

4

The Spectrum of Riemannium

The year: 1972. The scene: Afternoon tea in Fuld Hall at the Institute for Advanced Study in Princeton, New Jersey. The camera pans around the Common Room, passing by several Princetonians in tweeds and corduroys, then zooms in on Hugh Montgomery, boyish Midwestern number theorist with sideburns. He has just been introduced to Freeman Dyson, dapper British physicist.

Dyson: So tell me, Montgomery, what have you been up to?

Montgomery: Well, lately I've been looking into the distribution of the zeros of the Riemann zeta function.

Dyson: Yes? And?

Montgomery: It seems the two-point correlations go as . . . *(turning to write on a nearby blackboard)*:

$$1-\left(\frac{\sin(\pi x)}{\pi x}\right)^2.$$

Dyson: Extraordinary! Do you realize that's the pair-correlation function for the eigenvalues of a random Hermitian matrix? It's also a model of the energy levels in a heavy atomic nucleus—say, uranium-238.

I present this anecdote in cinematic form because I look forward to seeing it on the big screen someday. Besides, the screenplay genre gives me license to dramatize and embellish a little. By the time the movie opens at your local multiplex, the script doctors will have taken further liberties with the facts. For example, the equation for nuclear energy levels will have become the secret formula of the atomic bomb.

Even without Hollywood hyperbole, however, the chance encounter of Montgomery and Dyson was a genuinely dramatic moment. Their conversation revealed an unsuspected connection between areas of mathematics and physics that had seemed remote. Why should the same equation describe both the structure of an atomic nucleus and a mathematical sequence at the heart of number theory? And what do random matrices have to do with either of those realms? In the years since the meeting in Fuld Hall, the plot has thickened further, as random matrices have turned up in other unlikely places: games of solitaire, chaotic billiard tables, even the arrival times of buses in Cuernavaca, Mexico. Is it all just a cosmic coincidence, or is there something going on behind the scenes?

The Spectrum of Interstatium

How things distribute themselves in space or time or along some other dimension is a question that comes up in all the sciences. A biologist studies the distribution of genes along a chromosome; a seismologist records the temporal pattern of earthquakes; a mathematician ponders the sprinkling of prime numbers among the integers. The canonical example of such a distribution is an atomic spectrum—a series of bright, colored lines representing transitions between energy levels in an atom. Every chemical element has its own characteristic spectrum. The term *spectrum* is commonly applied to other discrete, one-dimensional distributions as well. Thus when Hugh Montgomery was examining the distribution of the zeros of the Riemann zeta function, we might say he was looking at the spectrum of a fanciful element called Riemannium.

Figure 4.1 shows samples of several such spectra, some of them mathematically defined and others derived from measurements. All the samples

Figure 4.1 One-dimensional distributions are analogous to atomic spectra, generated by imaginary chemical elements. The first three spectra are simple mathematical constructions: an array of evenly spaced lines (*periodium*), a random sequence (*aleatorium*), and a periodic array perturbed by a slight random jiggling of each level (*jigglium*). The next three all seem to share a similar texture: the nuclear energy spectrum of a real chemical element (*erbium*), the central 100 eigenvalues of a 300 by 300 random symmetric matrix (*eigenvalium*), and the heights of certain zeros of the Riemann zeta function (*Riemannium*). The remaining series are a miscellany: 100 consecutive prime numbers beginning with 103,613 (*primium*), the locations of 100 overpasses and underpasses along Interstate 85 (*interstatium*), positions of cross-ties on a railroad siding (*amtrakium*), growth rings in a fir tree on Mount St. Helens, Washington (*dendrochronomium*), and dates of 100 California earthquakes (*seismium*).

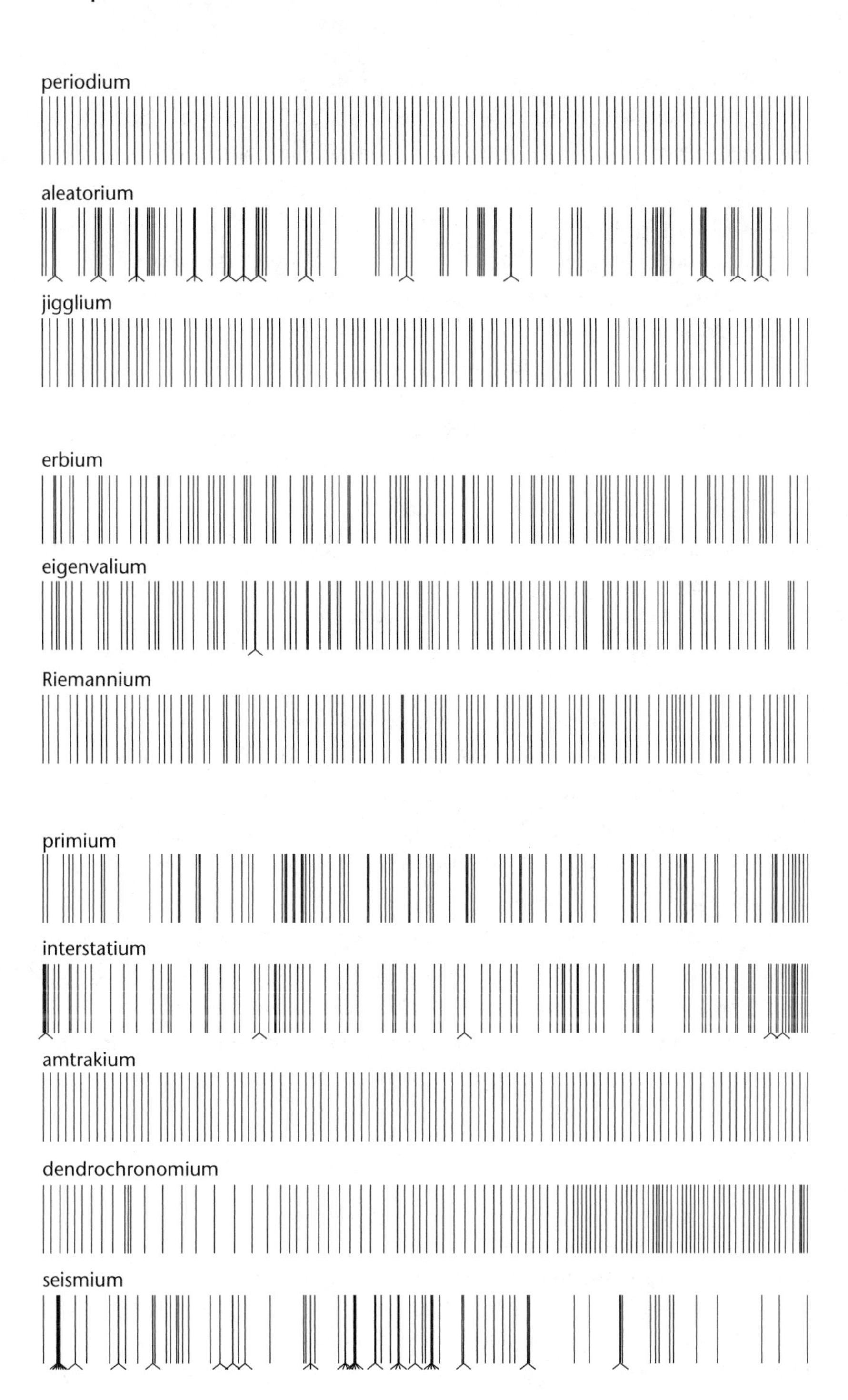
periodium
aleatorium
jigglium
erbium
eigenvalium
Riemannium
primium
interstatium
amtrakium
dendrochronomium
seismium

have been scaled so that exactly 100 levels fit in the space allotted. Thus the mean distance between levels is the same in all cases, but the patterns are nonetheless diverse. For example, the earthquake series (the spectrum of seismium) is highly clustered, which surely reflects some geophysical mechanism. The low-frequency fluctuations observed in tree-ring data (dendrochronomium) may have both biological and climatological causes. And it's anyone's guess how to explain the locations of bridges recorded while driving along a stretch of interstate highway (interstatium).

The simplest of all distributions is a periodic one. Think of a picket fence or the monotonous ticking of a clock. All the intervals between elements of the series are exactly the same. The obvious counterpoint to such a repetitive pattern is a totally random one. And between these extremes of order and disorder there are various intermediate possibilities, such as a jiggled picket fence, where periodic levels have each been randomly displaced by a small amount.

The human eye is quick to detect the differing textures of such patterns, but a little math can make the distinctions clearer. There is seldom much hope of predicting the positions of individual elements in a series. The aim is statistical understanding—a description of a typical pattern rather than a specific one. Two useful statistical measures are the nearest-neighbor spacings and the two-point correlation function (see figure 4.2).

To calculate the nearest-neighbor spacings, measure the distance from each level to the next level, sort these numbers into intervals or bins of appropriate width, then plot a histogram of the frequencies—the number of distances in each bin. Applied to the periodic distribution, the graph is not much to look at; all the spacings are the same, and so they all fall into a single bin. The nearest-neighbor spectrum of the random distribution is more interesting; the smallest spacings are the most common, and the frequencies decline exponentially for larger distances. The jiggled

Figure 4.2 The nearest-neighbor and pair correlation functions summarize the statistics of a spectrum. For a periodic series the nearest-neighbor curve has a single nonzero point; for random levels the curve follows an exponential law; the jiggled series yields a graph that looks Gaussian. The pair correlation function measures the number of levels separated by any given distance. For a periodic series the function has a sharp peak at each multiple of the nearest-neighbor spacing; for a random series the function is flat (apart from random noise); the jiggled series yields a sequence of humps.

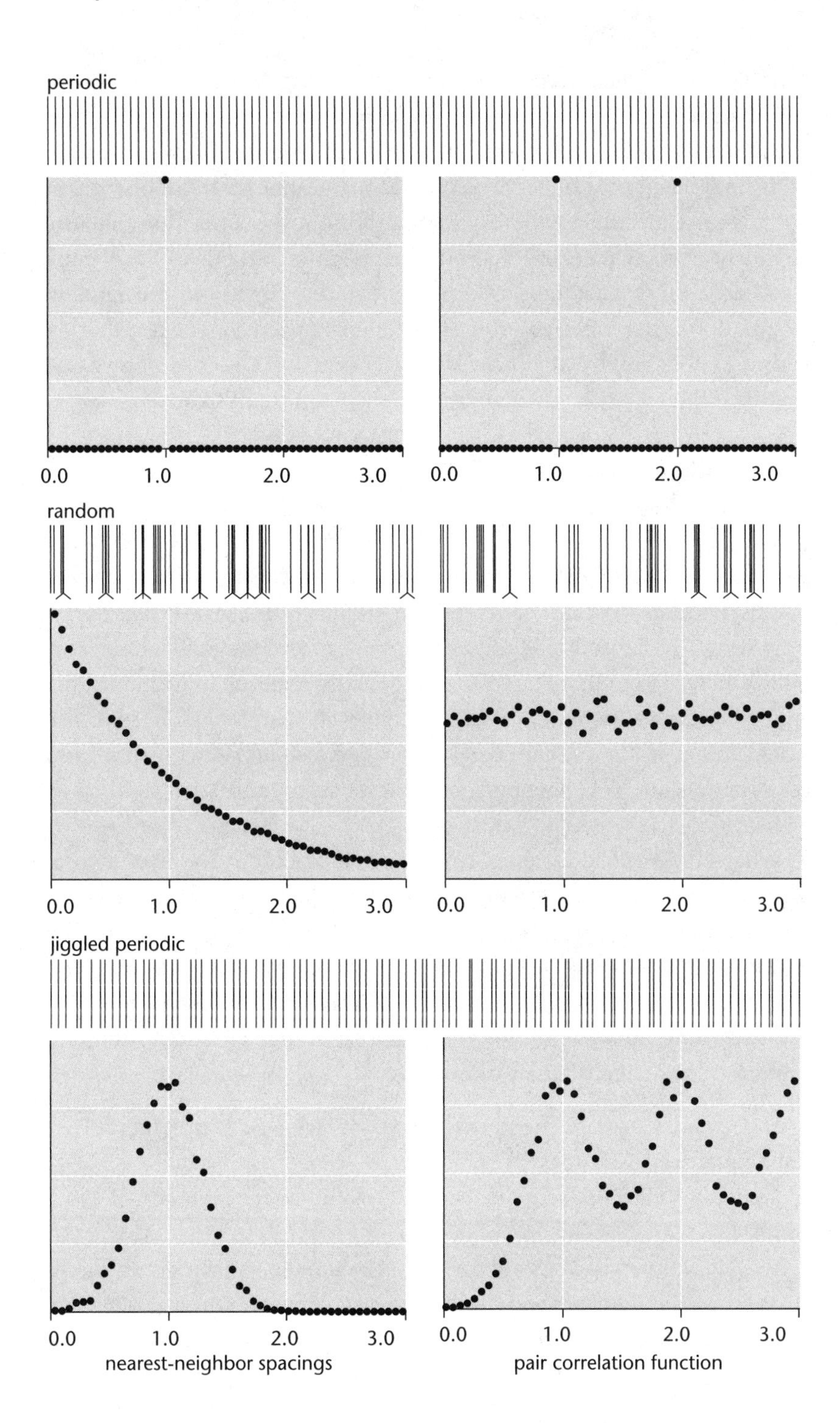
periodic
0.0
1.0
2.0
3.0
0.0
1.0
2.0
3.0
random
0.0
1.0
2.0
3.0
0.0
1.0
2.0
3.0
jiggled periodic
0.0
1.0
2.0
3.0
0.0
1.0
2.0
3.0
nearest-neighbor spacings
pair correlation function

periodic pattern yields a bell-shaped curve centered on the original periodic distance.

The pair correlation function mentioned by Montgomery and Dyson captures some of the same information as the nearest-neighbor histogram, but it is calculated differently. For each distance x, the correlation function counts how many pairs of levels are separated by x, whether or not those levels are nearest neighbors. The pair correlation function for a random distribution is flat, since all intervals are equally likely. As the distribution becomes more orderly, the pair correlation function develops humps and ripples; for the periodic distribution it is a series of sharp spikes.

Erbium and Eigenvalium

Among the sample spectra in figure 4.1 is a series of 100 energy levels of an atomic nucleus, measured with great finesse by H. I. Liou and James Rainwater and their colleagues at Columbia University. The nucleus in question is that of the rare earth element erbium-166. A glance at the spectrum reveals no obvious patterns or regularities; nevertheless, the texture is quite different from that of a purely random distribution. In particular, the erbium spectrum has fewer closely spaced levels than a random sequence would. It's as if the nuclear energy levels come equipped with springs to keep them apart. This phenomenon of level repulsion is characteristic of all heavy nuclei.

What kind of mathematical structure could account for such a spectrum? This is where those eigenvalues of random matrices enter the picture. They were proposed for this purpose in the 1950s by the physicist Eugene P. Wigner. As it happens, Wigner was another Princetonian, who can therefore make an appearance in our movie. Let him be the kindly professor who explains things to a hapless journalist who is struggling to make sense of the story. The dialogue might go like this:

Wigner: So, you want me to tell you about eigenvalues of random Hermitian matrices?

Reporter: I could use some help. I don't really understand any of those words.

Wigner: Not to worry. As my friend von Neumann used to say, in mathematics you don't understand things; you just get used to them. For starters all you need to know is that a matrix is a table of numbers, arranged in columns and rows. We're only going to talk about square matrices, with the same number of columns and rows. Okay?

Reporter: So far so good.

Wigner: Now for the random part. What can I say? Pick some numbers. Roll the dice, spin the roulette wheel, open the telephone book. It doesn't matter where they come from.

Reporter: What about "Hermitian"?

Wigner: Right. Now I take a step backwards. Let's first build something simpler, a real symmetric matrix. Let the main diagonal of the matrix—from the upper left corner to the lower right corner—act as a kind of mirror, so that all the elements above and to the right of the diagonal are reflected below and to the left. We take some numbers—just ordinary real numbers, the kind you use every day—11, 14.16, π—and fill the upper triangle with them. Then we copy the same numbers into the mirror image cells in the lower triangle.

Reporter: But then the matrix isn't random anymore, is it?

Wigner: Yeah, that's one of things you just get used to. Anyway, now we have our real symmetric matrix. The Hermitian matrix is just a slight variation. Instead of real numbers we take complex numbers. You know about them?

Reporter: I think so. Real part and imaginary part?

Wigner: Exactly—numbers of the form $a+bi$, where i is the square root of minus 1. So we fill the upper triangle with complex numbers, and then there's one more little trick. We don't just copy them to the lower triangle; instead we fill in the complex conjugate of each number. If it's $a+bi$ up top, then it's $a-bi$ down below, and vice versa. Any questions?

Reporter: Why is it called Hermitian?

Wigner: After Charles Hermite, French mathematician.

Reporter: And eigenvalues? Where do they come from?

Wigner: Ah, calculating eigenvalues is easy. If you have a square matrix M, you start up Matlab and type `eig(M)`!

At this point, as director of the film, I must stand up and yell "Cut!" Although Wigner's pithy advice on calculating eigenvalues is sound enough, it's a flagrant anachronism. There was no Matlab in 1972. Furthermore, what's needed here is not so much a recipe for calculating eigenvalues as a hint about what they are.

Every $n \times n$ matrix corresponds to an nth-degree polynomial equation, with coefficients determined by the elements of the matrix. The eigenvalues are the roots of this equation. There are n of them. In general, the eigenvalues are complex numbers, even if the matrix elements are real. However, symmetry has an effect here. For a real symmetric matrix all the eigenvalues are real numbers (see figure 4.3). What's more surprising, this is also true

–0.0686	0.3109	–0.3535	–0.1896	–0.2760	–1.1741
0.3109	–0.3934	–0.0913	–0.7215	–0.1127	0.1722
–0.3535	–0.0913	1.1197	1.0075	–0.4051	–0.6815
–0.1896	–0.7215	1.0075	0.6768	–0.6676	0.3700
–0.2760	–0.1127	–0.4051	–0.6676	0.4609	0.1283
–1.1741	0.1722	–0.6815	0.3700	0.1283	–1.9240

characteristic polynomial:
$x^6 + 0.1285x^5 - 7.4107x^4 + 1.6865x^3 + 4.9804x^2 - 1.2215x - 0.2040 = 0$

eigenvalues:
–2.7699, –0.8340, –0.1157, 0.3840, 0.8398, 2.3673

Figure 4.3 A random symmetric matrix is a square array of randomly chosen numbers with a distinctive symmetry. The main diagonal *(white squares extending from upper left to lower right)* is a mirror axis, and entries in the upper right triangle are reflected in the lower left one. The shading of the cells visually emphasizes this symmetry. The 6×6 matrix has a sixth-degree characteristic polynomial equation, with coefficient determined by the elements of the matrix. The six roots of the equation are the eigenvalues of the matrix.

of a Hermitian matrix; even though the matrix elements are complex, the eigenvalues are real. Because they are real numbers, they can be sorted from smallest to largest and arranged along a line. In this configuration they look a lot like the energy spectrum of a heavy nucleus. Of course, the eigenvalues of a random matrix do not match any specific nuclear spectrum level for level, but statistically the resemblance is strong. In particular, they exhibit level repulsion; closely spaced levels are rarer than they would be in a random spectrum.

When I first heard of the random matrix conjecture in nuclear physics, what surprised me most was not that it might be true but that anyone would ever have stumbled on it. But Wigner's idea was not just a wild guess. The key idea was already implicit in Werner Heisenberg's formulation of quantum mechanics, where the internal state of an atom is represented by a matrix whose eigenvalues are the energy levels of the atomic spectrum. Each atom is associated with a specific matrix, which determines the exact energy levels. What Wigner recognized is that the statistical qualities of the spectrum are not terribly sensitive to the specific matrix elements. In a sense, almost any large, symmetric matrix will do. Thus a random matrix is not one with some special or unusual property; on the contrary, it's just a typical matrix.

Eulerium and Riemannium

So much for nuclear physics. What about number theory and the Riemann zeta function?

The most celebrated sequence in number theory is that of the primes, the numbers divisible only by 1 and themselves: 2, 3, 5, 7, 11 . . . . The overall trend in this series is well known. In the neighborhood of any large integer x, the proportion of numbers that are prime is approximately $1/\log x$, which implies that although the primes go on forever, they get sparser as you move farther out on the number line. Superimposed on this gradual thinning of the crop are smaller-scale fluctuations that are harder to understand in detail. The sequence of primes looks quite random and erratic, and yet it cannot possibly have the same nearest-neighbor statistics as a truly random spectrum. The nearest that two primes can approach each other (except in one anomalous case) is 2. Pairs that have this minimum spacing, such as 29 and 31, are called twin primes. No one knows whether there are infinitely many of them.

In addition to directly exploring the primes, mathematicians have taken a roundabout approach to understanding their distribution by way of the Riemann zeta, or ζ, function. This function, although named both by and for Bernhard Riemann, was first studied in the eighteenth century by Leonhard Euler, who defined it as a sum over all the natural numbers:

$$\zeta(s) = \sum_{n=1}^{\infty} \frac{1}{n^s}.$$

In other words, take each whole number n from 1 to infinity, raise it to the power s, take the reciprocal, and add up the entire series. The sum is finite whenever s is greater than 1. For example, the zeta series for $s = 2$ begins

$$\zeta(2) = \frac{1}{1^2} + \frac{1}{2^2} + \frac{1}{3^2} + \cdots \quad = \frac{1}{1} + \frac{1}{4} + \frac{1}{9} + \cdots .$$

Euler showed that this series of infinitely many terms converges to a finite sum, namely, $\pi^2/6$, or about 1.645.

Euler also proved a remarkable identity, equating the summation formula, with its one term for each natural number, to a product formula that has one term for each prime number. This second definition states:

$$\zeta(s) = \prod_{p \text{ prime}} \frac{1}{1 - \frac{1}{p^s}}.$$

The recipe in this case is to take each prime p from 2 to infinity, raise it to the power s, then after some further arithmetic multiply together the terms for all values of p. Again taking the example of $s = 2$, the first few factors of the product evaluate to

$$\zeta(2) = \frac{4}{3} \cdot \frac{9}{8} \cdot \frac{25}{24} \cdot \frac{49}{48} \cdots .$$

As with the sum, this infinite product yields a finite result; indeed, it is the same as that of the summation, $\pi^2/6$. This connection between a sum over all integers and a product over all primes was a hint that the zeta function might have something to say about the distribution of primes among the integers. In fact the two series are intimately related.

Riemann's contribution, in 1859, was to extend the domain of the zeta function. In Euler's time, the variable s was taken to range over the positive integers. Later, Pafnuty Chebyshev showed that the function is well behaved not just for integers but for all real values of s greater than 1. Riemann found a clever way of defining the function for all complex numbers with the single exception of $s = 1$ (because $\zeta(1)$ is equal to $\frac{1}{1} + \frac{1}{2} + \frac{1}{3} + \frac{1}{4} + \cdots$, a series that diverges to infinity).

Whereas the real numbers live on a number line, extending from $-\infty$ to $+\infty$, the complex numbers fill an entire plane, with real and imaginary

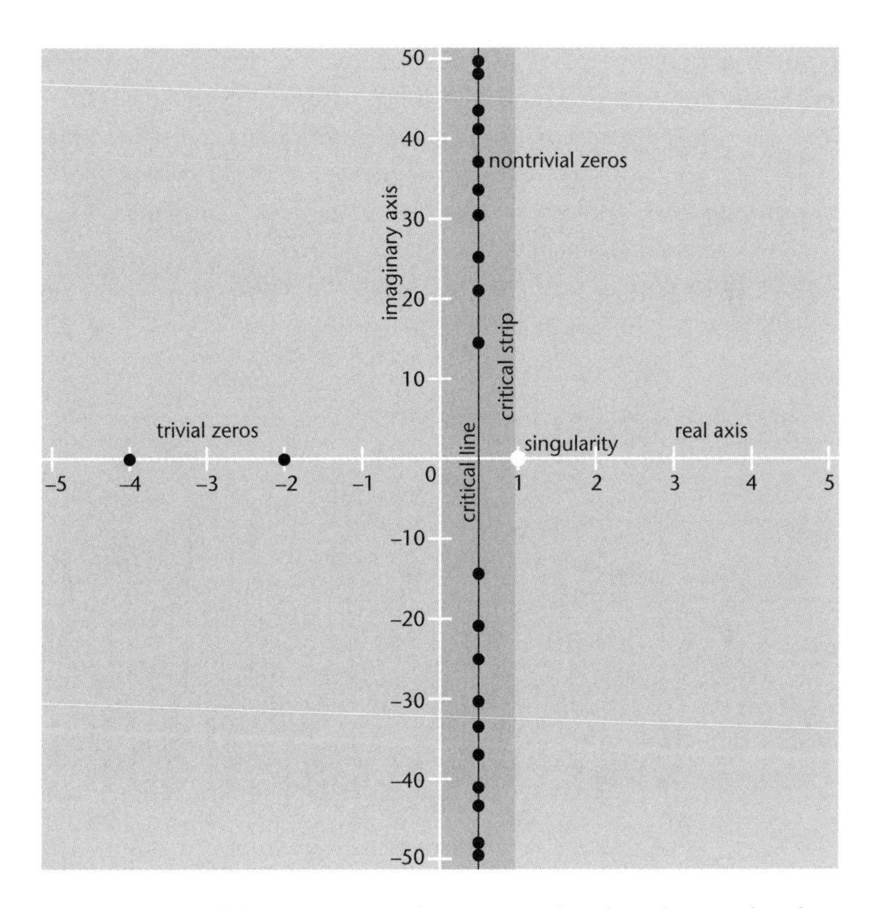

Figure 4.4 Zeros of the Riemann zeta function are plotted on the complex plane. There are "trivial" zeros on the negative real axis, but the zeros that attract the interest of number theorists lie in a critical strip parallel to the imaginary axis. The Riemann hypothesis predicts that all these zeros lie on the critical line in the middle of the strip, with real coordinate 1/2. The distribution of the heights of the zeros above and below the real axis encodes information about the distribution of prime numbers.

axes. Over much of the complex plane the zeta function turns out to be wildly oscillatory, crossing from positive to negative values infinitely often. The crossing points, where $\zeta(s) = 0$, are called the zeros of the zeta function (see figure 4.4). There is an infinite series of them along the negative real axis, but these are not looked upon with great interest. Riemann called attention to a different infinite series of zeros lying above and below the real axis in a vertical strip of the complex plane that includes all numbers whose real part is between 0 and 1. Riemann calculated the locations of the first three of these zeros and found that they lie right in the middle of the strip, on the critical line with real coordinate 1/2. On the basis of this evidence, plus incredible intuition, he conjectured that all the complex zeros are on the critical line. This is the Riemann hypothesis—still unproved and widely regarded as the juiciest prize plum in all of contemporary mathematics.

In the years since Riemann located the first three zeta zeros, quite a few more have been found. A cooperative computing network called ZetaGrid,

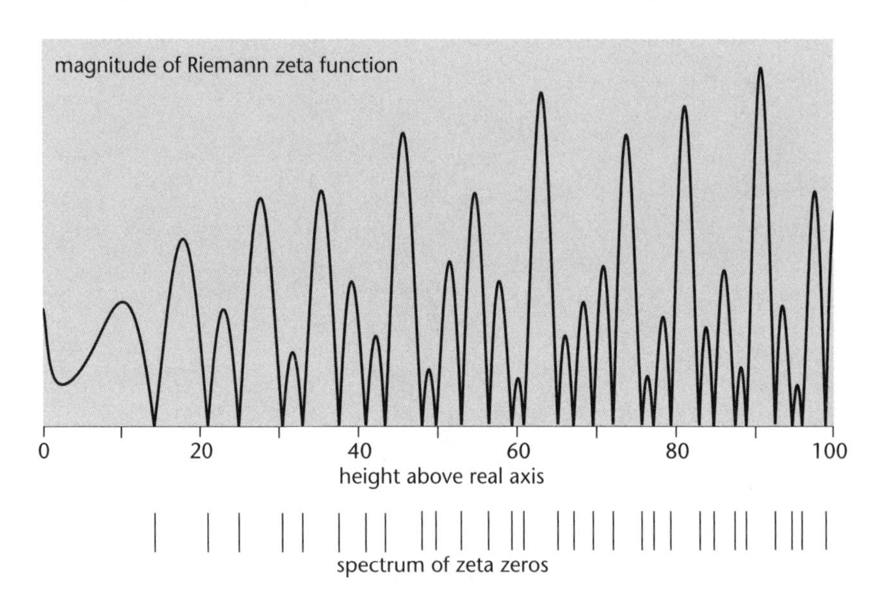

Figure 4.5 Magnitude of the Riemann zeta function is plotted along the critical line up to a height of 100 above the real axis. In that region the function has 29 zeros—too few for a quantitative analysis of their distribution, but some basic features of the pattern are already apparent. The zeros grow denser with increasing height, and there's a hint of level repulsion.

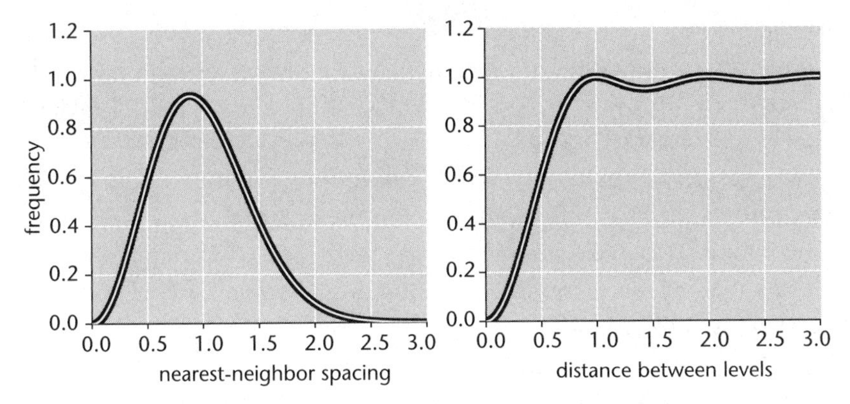

Figure 4.6 Nearest-neighbor spacings and pair correlation function for zeros of the Riemann zeta function closely match the predictions of random matrix theory. Heavy black lines represent the positions of a billion zeta zeros above the 10^{23}rd zero; the thin white line is the predicted spacing. Note the evidence of level repulsion; in both graphs closely spaced zeros are rare. The data were supplied by Andrew M. Odlyzko.

organized by Sebastian Wedeniwski, checked a few hundred billion zeros; then in 2004 Xavier Gourdon leaped ahead to 10 trillion zeros. Every one is on the critical line. There's even a proof that infinitely many zeros lie on the line, but what's wanted is a proof that none lie anywhere else. That goal remains out of reach.

In the meantime, other aspects of the zeta zeros have come under scrutiny. Assuming that all the zeros are indeed on the critical line, what is their distribution along that line? How does their density vary as a function of the height, T, above or below the real axis?

As with the primes, the overall trend in the abundance of zeta zeros is known. The trend goes the opposite way. Whereas primes get rarer as they get larger, the zeta zeros crowd together with increasing height (see figure 4.5). The number of zeros in the neighborhood of height T is proportional to log T, signifying a slow increase. But, again, the trend is not smooth, and the details of the fluctuations are all-important. Gaps and clumps in the series of zeta zeros encode information about corresponding features in the sequence of primes.

Montgomery's work on the pair correlation function of the zeta zeros was a major step toward understanding the statistics of the fluctuations. And the encounter in Fuld Hall, when it emerged that Montgomery's correlation formula is the same as that for eigenvalues of random matrices, ignited further interest. The correlation function implies level repulsion among the zeros just as it does in the nucleus, producing a deficiency of closely spaced zeros.

Montgomery's result is not a theorem; his proof of it is contingent on the truth of the Riemann hypothesis. But the accuracy of the correlation function can be tested by comparing the theoretical prediction with computed values of zeta zeros. Over a period of more than 20 years, Andrew M. Odlyzko took the computation of zeta zeros to heroic heights in order to perform such tests. For this purpose it is not enough to verify that the zeros lie on the critical line; the program must accurately measure the height of each zero along that line, which is a more demanding task. One of Odlyzko's early papers was titled "The 10^{20}th zero of the Riemann zeta function and 175 million of its neighbors." Since then he has gone on to compute even longer series of consecutive zeros at even greater heights, reaching the neighborhood of the 10^{23}rd zero. The agreement between predicted and measured correlations is striking, and it gets better and better with increasing height (see figure 4.6.).

Permutium and Omnibusium

The mania for random matrices began with nuclear physics and number theory, but it has spread to many other areas. One example is combinatorics. Take a sequence of numbers, say, 1–10, and randomly rearrange them, yielding a permutation such as 10, 3, 8, 4, 7, 5, 1, 6, 9, 2. Now find the longest subsequence of increasing numbers (which need not be consecutive). In this case it is 3, 4, 5, 6, 9. When this experiment is repeated many times (and with longer sequences), the lengths of the longest subsequences have the same distribution as the largest eigenvalues of random Hermitian matrices. The connection between permutations and random matrices was discovered in 1999 by Jinho Baik, Percy Deift, and Kurt Johansson. Soon after, David Aldous and Persi Diaconis applied the same kind of analysis in a different context: a solitaire card game called patience sorting. In this case the largest eigenvalues predict the number of piles of cards at the end of the game.

The results on permutations and patience sorting differ in a subtle way from much of the earlier work. Nuclear spectra and the heights of zeta zeros are modeled by an entire series of matrix eigenvalues, but the combinatorial work focuses on the margins of the distribution—just the largest eigenvalue or the first few eigenvalues. This marginal distribution is now called the Tracy-Widom distribution, after Craig Tracy and Harold Widom. It has found dozens of new applications, such as models of organic growth and the patterns formed when tiles are placed at random.

A field called quantum chaos has also been deeply influenced by random matrix theory. Imagine a frictionless billiard table shaped like a football stadium. If you set a ball rolling across the surface and bouncing off the cushions at the edge of the table, it may follow a stable trajectory, retracing the same finite path over and over. But the slightest adjustment to the initial direction can produce chaotic motion, with the ball never repeating the same combination of position and velocity. The billiard system can be seen as a weird kind of atom, where the ball confined to the surface of the table is analogous to an electron orbiting the nucleus. In quantum mechanics, any atom—even a weird one—has a spectrum of discrete energy levels, but the possibility of chaos changes the nature of that spectrum. Oriol Bohigas and Marie-Joya Giannoni discovered that the Wigner distribution for random matrices is a good match to the billiard spectrum.

One final example brings the recondite subject of random matrices a little closer to everyday life. In 2000 Milan Krbálek and Petr Šeba studied the arrival times of buses at various stops along a route in the city of Cuernavaca, Mexico. Other studies of public transit have shown that buses often form platoons. The lead bus picks up all the passengers, causing it to fall behind schedule, while the following buses make fewer stops and therefore catch up. In Cuernavaca the buses are owned by the drivers, whose income depends on the fares collected. The drivers have therefore devised schemes to optimize the intervals between buses, deliberately slowing down if another bus has recently passed. The result is yet another spectrum exhibiting level repulsion. Krbálek and Šeba found that the intervals are modeled well by the eigenvalues of random matrices. The analysis was later made more precise by Jinho Baik, Alexei Borodin, Percy Deift, and Toufic Suidan.

The Operator of the Universe

Is it all just a fluke, this apparent link between matrix eigenvalues, nuclear physics, zeta zeros, and Mexican buses? It could be, although a universe with such chance coincidences in its fabric might be considered even stranger than one with mysterious causal connections.

A likelier explanation is that the statistical distributions seen in these cases are simply very common ways for things to organize themselves. There is an analogy with the Gaussian distribution, which turns up everywhere in nature because many different processes all lead to it. When multiple, random, independent contributions are summed up, the outcome is often the familiar bell-shaped Gaussian curve. Maybe some similar principle makes the eigenvalue distribution ubiquitous. Thus for Montgomery and Dyson to come up with the same correlation function would not be such a long shot after all.

Still another view is that the zeros of the zeta function really do represent a spectrum—a series of energy levels just like those of the erbium nucleus but generated by the mathematical element Riemannium. This idea traces back to David Hilbert and George Pólya, who both suggested (independently) that the zeros of the zeta function might be the eigenvalues of some unknown Hermitian operator. An operator is a mathematical concept that seems on first acquaintance rather different from a matrix—it is a function that applies to functions—but operators, too, have eigenvalues, and a Hermitian operator has symmetries that make all the eigenvalues real numbers, just as in the case of a Hermitian matrix.

If the Hilbert-Pólya thesis is correct, then random matrix methods succeed in number theory for essentially the same reason they work in nuclear physics—because the detailed structure of a large matrix (or operator) is less important than its global symmetries, so that any typical matrix with the right symmetries will produce statistically similar results. Behind these approximations stands some unique Hermitian operator, which determines the exact position of all the Riemann zeros and hence the distribution of the primes.

Is that universal operator really out there, waiting to be discovered? Will it ever be identified? For the answers to those questions you'll have to wait for the movie. I'll be giving no spoilers here.

Acknowledgments

I owe the title of this essay to Oriol Bohigas and Patricio Leboeuf, who mentioned "the spectrum of Riemannium" in 1999, during a semester-long program on random matrices at the Mathematical Sciences Research Institute in Berkeley. (With A. G. Monastra they later published a paper titled "The Riemannium.") In my long (and ongoing) struggle to understand random matrices and the Riemann zeta function I have also benefited from discussions with Alan Edelman, Brian Conrey, Persi Diaconis, David Farmer, Peter Forrester, Andrew Odlyzko, Peter Sarnak, Craig Tracy, and others.

5

Unwed Numbers

When the Sudoku craze first swept through the English-speaking world in the early 2000s, many newspapers accompanied the puzzles with the reassuring slogan "No mathematics required." Apparently the publishers worried that readers accustomed to crosswords and other lexical puzzles might be put off by grids filled with numbers.

What "no math required" really meant was "no arithmetic required." You don't have to add up columns of figures; you don't even have to count. As a matter of fact, the symbols in the grid need not be numbers at all; letters or colors or fruits would do as well. Thus it's true that solving the puzzle is not a test of skill in arithmetic. On the other hand, when you look into Sudoku a little more deeply, you'll find plenty of mathematical ideas lurking in the background.

A Puzzling Provenance

The standard Sudoku puzzle grid has 81 cells, organized into nine rows and nine columns and also marked off into nine 3×3 blocks. Some of the cells are already filled in with numbers called givens. The aim is to complete the grid in such a way that every row, every column, and every block has exactly one instance of each number from 1 to 9 (see figure 5.1). A well-formed puzzle has one and only one solution.

The name Sudoku is Japanese, but the game itself is almost surely an American invention. The earliest known examples were published in 1979 in *Dell Pencil Puzzles & Word Games*, where they were given the title Number Place. The constructor of the puzzles is not identified in the magazine, but Will Shortz, the puzzles editor of the *New York Times*, thinks he has identified the author through a process of logical deduction reminiscent of what it takes to solve a Sudoku. Shortz examined the list of contributors in several Dell magazines; he found a single name that was always present if an issue included a Number Place puzzle and never present otherwise. The putative inventor identified in this way was Howard Garns, an architect from Indianapolis who died in 1989. Mark Lagasse, senior executive editor of Dell Puzzle Magazines, concurs with Shortz's conclusion, although he says Dell has no records attesting to Garns's authorship, and none of the editors now on the staff were there in 1979.

The later history is easier to trace. Dell continued publishing the puzzles, and in 1984 the Japanese firm Nikoli began including puzzles of the same design in one of its magazines. (Puzzle publishers, it seems, are adept at the sincerest form of flattery.) Nikoli named the puzzle "su–ji wa dokushin ni kagiru," (数字は独身に限る), which I am told means "the numbers must be single"—single in the sense of unmarried. The name was soon shortened to Sudoku (数独), which is usually translated as "single numbers." Nikoli secured a trademark on this term in Japan, and so later Japanese practitioners of sincere flattery have had to adopt other names. Ed Pegg, writing in the Mathematical Association of America's *MAA Online*, points out an ironic consequence. Many Japanese know the puzzle by its English name, Number Place, whereas the English-speaking world prefers the Japanese name Sudoku.

The next stage in the puzzle's east-to-west circumnavigation was a brief detour to the Southern Hemisphere. Wayne Gould, a New Zealander who had been a judge in Hong Kong before the British lease expired in 1997, discovered Sudoku on a trip to Japan and wrote a computer program to generate the puzzles. Eventually he persuaded the *Times* of London to print them; the first appeared in November 2004. The subsequent fad in the United Kingdom was swift and intense. Other newspapers joined in, with the *Daily Telegraph* running the puzzle on its front page. There was boasting about who had the most and the best Sudokus, and bickering over the supposed virtues of handmade versus computer-generated puzzles. In July 2005 a Sudoku tournament was televised in Britain; the event was promoted by carving a 275-foot grid into a grassy hillside near Bristol. (It soon emerged that this "world's largest Sudoku" was defective.)

Sudoku came back to the United States in the spring of 2005. Here, too, the puzzle became a popular pastime, although perhaps not quite the all-consuming obsession it was in Britain. There was no dramatic dip in the U.S. gross domestic product as a result of mass distraction. On the other hand, I must confess that my own motive for writing on the subject is partly to justify the appalling number of hours I have squandered solving Sudokus.

Although the history of Sudoku apparently began with Howard Garns in the 1970s, the puzzle also has an interesting prehistory. In the 1950s W. U. Behrens of Hannover, Germany, was working to improve the design of agricultural experiments. If several seed varieties or several pesticide treatments are to be tested in a single field, they should be distributed across the area

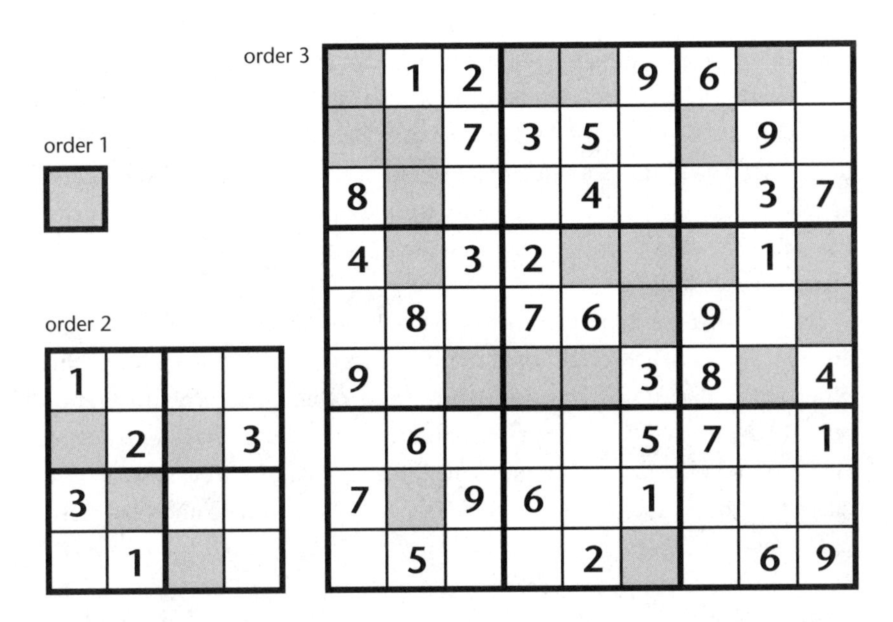

Figure 5.1 Sudoku puzzles have to be filled in so that each number appears exactly once in each column, each row, and each of the blocks delineated by heavier lines. The order 1 puzzle is a trivial 1×1 grid; the order 2 Sudoku is a 4×4 grid to be filled with the integers from 1 to 4; the order 3 puzzle is a 9×9 grid where the allowed numbers are 1 through 9. Some useful terminology: The individual compartments are *cells*; the $n \times n$ groups of cells are *blocks*; the cells are arranged in horizontal *rows* and vertical *columns*; the blocks are organized in horizontal *bands* and vertical *stacks*; the union of a cell's row, column, and block is called its *neighborhood*; the numbers supplied in the initial state are *givens*. The order 3 Sudoku shown here is a variation on the first puzzle published, in 1979, in *Dell Pencil Puzzles & Word Games*; by present-day standards it is quite easy. Shaded cells are fully determined by the givens alone.

	5		1	
1		4		3
5			2	4
		5	3	
3				

in such a way that varying conditions in different plots won't bias the outcome of the trial. Behrens arranged the plots in columns, rows, and blocks, with each seed variety or treatment represented exactly once in each of these neighborhoods. He called the layouts *gerechte Anordnungen*, which might be translated as "fair arrangements." When Sudoku became popular, R. A. Bailey, a British statistician who had earlier evaluated Behrens's proposal, pointed out the similarity between the puzzle and the experimental designs. (Together with Peter J. Cameron and Robert Connelly, Bailey published a full analysis in 2008.)

Some of the Behrens designs correspond exactly to 4×4 and 9×9 Sudokus, with square blocks of 2×2 or 3×3 cells. But these arrangements work only when the block size is a perfect square. For other sizes, Behrens came up with some pleasant arrangements with nonsquare blocks, such as the five-cell array above. I have erased some of his numerical entries to turn it into a pseudo-Sudoku.

Even earlier Sudoku precedents have been discovered by Christian Boyer, a specialist in recreational mathematics. In the last decades of the nineteenth century, several newspapers and other periodicals in France published number grid puzzles that have some of the characteristics of Sudokus. No one example exhibits all the elements of a true Sudoku, but taken together these number puzzles from *la Belle Époque* come very close to the modern game.

Hints and Heuristics

If you take a pencil to a few Sudoku problems, you'll quickly discover various useful rules and tricks. The most elementary strategy for solving the puzzle is to examine each cell and list all its possible occupants—that is, all the numbers not ruled out by a conflict with another cell. If you find a cell that has only one allowed value, then obviously you can write in that value. The complementary approach is to note all the cells within a row, a column, or a block where some particular number can appear; again, if there is a number that can occupy only one position, then you should put it there. In either case, you can eliminate the selected number as a candidate in all other cells in the same neighborhood.

Some Sudokus can be solved by nothing more than repeated application of these two rules—but if all the puzzles were so straightforward, the fad would not have lasted long. Barry Cipra, a mathematician and writer in Northfield, Minnesota, describes a hierarchy of rules of increasing complexity. The two rules just mentioned constitute level 1; they restrict a cell to a single value or restrict a value to a single cell. At level 2 are rules that apply to pairs of cells within a row, column, or block; when two such cells have only two possible values, those values are excluded elsewhere in the neighborhood. Level 3 rules work with triples of cells and values in the same way. In principle, the tower of rules might rise all the way to level 9.

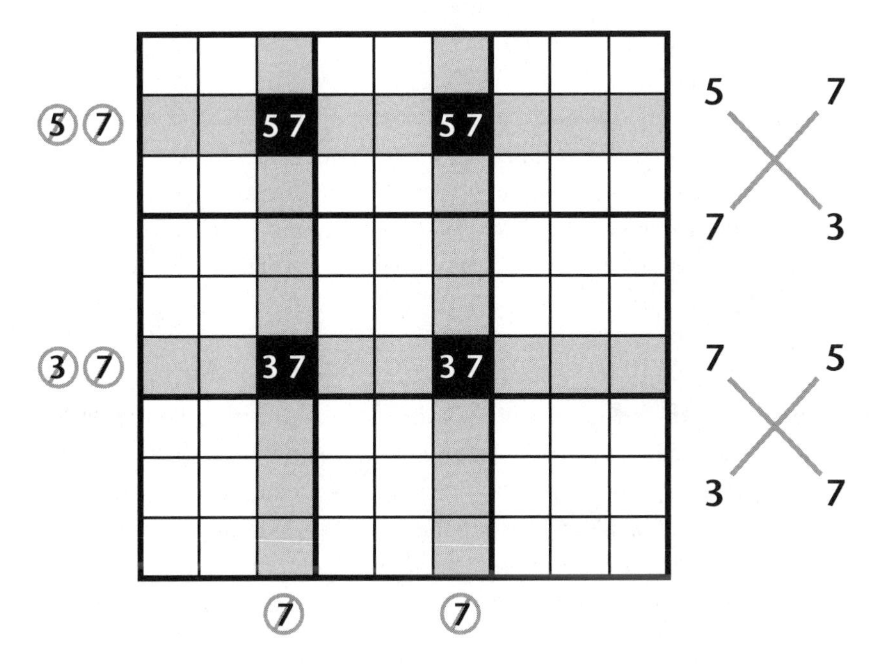

Figure 5.2 The x-wing motif is one of several higher-level Sudoku solving aids. Assume that the two black cells in the second row are constrained to hold either 5 or 7; it follows immediately that 5 and 7 are ruled out as candidates for all the other cells in row 2. Similarly, in row 6, knowing that 3 and 7 must occupy the black squares excludes them from consideration elsewhere in the row. The x-wing strategy adds further information: 7s are barred from the gray cells in the two columns containing the black cells. The diagrams at right explain why. There are only two valid ways to arrange the digits in the black cells, and in both cases a 7 appears in each column. (The diagrams also explain the source of the name x-wing.)

Puzzle enthusiasts have developed a number of colorfully named tricks for dealing with specific configurations: x-wing, swordfish, jellyfish. The x-wing, for example, is a level 4 technique in Cipra's hierarchy; it exploits constraints on four cells at the corners of a rectangle to eliminate candidates elsewhere (see figure 5.2). Beyond such ad hoc techniques, several authors have formulated comprehensive guides to solution strategies; works by J. F. Crook, by David Eppstein, and by Tom Davis are particularly impressive. Also, Zhe Chen sketches a "heuristic reasoning" procedure and claims that it can solve all known Sudokus.

When you are solving a specific puzzle, the search for patterns that trigger the various rules is where the fun is. But if you are trying to gain a higher-level understanding of Sudoku, compiling a catalog of such techniques doesn't seem very promising. The rules are too many, too various, and too specialized.

Rather than discuss methods for solving specific puzzles, I want to explore some more general questions about Sudoku, looking at it as a computational problem as well as a logic puzzle. How hard a problem is it? Pencil-and-paper experience suggests that some instances are much tougher than others, but are there any clear-cut criteria for ranking or classifying the puzzles?

Counting Solutions

In the search for general principles, a first step is to generalize the puzzle itself. The standard 81-cell Sudoku grid is not the only possibility. For any positive integer n, we can draw an order n Sudoku grid with n^2 rows, n^2 columns, and n^2 blocks; the grid has a total of n^4 cells, which are to be filled with numbers in the range from 1 to n^2. The standard grid with 81 cells is of order 3. Some publishers produce puzzles of order 4 (256 cells) and order 5 (625 cells). On the smaller side, there's not much to say about the order 1 puzzle. The order 2 Sudoku (with four rows, columns, and blocks, and 16 cells in all) is no challenge as a puzzle, but it does serve as a useful test case for studying concepts and algorithms.

How many Sudoku solutions exist for each n? To put the question another way: Starting from a blank grid—with no givens at all—how many ways can the pattern be completed while obeying the Sudoku constraints? As a first approximation, we can simplify the problem by ignoring the blocks in the

Sudoku grid, allowing any solution in which each column and each row has exactly one instance of each number. A pattern of this kind is known as a Latin square, and it was already familiar to Leonhard Euler more than 200 years ago.

Consider the 4×4 Latin square (which corresponds to the order 2 Sudoku). Euler counted the ways of arranging the numbers 1, 2, 3, and 4 in a square array with no duplications in any row or column. There are exactly 576. It follows that 576 is an upper limit on the number of order-2 Sudokus. This is an upper limit because every Sudoku solution is necessarily a Latin square, but not every Latin square is a valid Sudoku. In a series of postings in 2005 on a (now defunct) Sudoku programmers forum, Frazer Jarvis of the University of Sheffield showed that exactly half the 4×4 Latin squares are Sudoku solutions; that is, there are 288 valid arrangements.

Moving to higher-order Sudokus and larger Latin squares, the counting gets harder in a hurry. Euler got only as far as the 5×5 Latin square, and the 9×9 Latin squares were not enumerated until 1975; the tally for the latter is 5,524,751,496,156,892,842,531,225,600, or about 6×10^{27}. The order 3 Sudokus must be a subset of these squares. They were counted in 2005 by Bertram Felgenhauer of the Technical University of Dresden in collaboration with Jarvis. The total they computed is 6,670,903,752,021,072,936,960, or 7×10^{21}. Thus, among all the 9×9 Latin squares, a little more than 1 in a million are also Sudoku grids.

It's a matter of definition, however, whether all those patterns are really different. The Sudoku grid has many symmetries. If you take any solution and rotate it by a multiple of 90 degrees, you get another valid grid; in the preceding tabulations, these variants are counted as separate entries. Beyond the obvious rotations and reflections, you can permute the rows within a horizontal band of blocks or the columns within a vertical stack of blocks, and you can also freely shuffle the bands and stacks themselves. Furthermore, the numerals in the cells are arbitrary tokens, which can also be permuted; for example, if you switch all the fives and sixes in a puzzle, you get another valid puzzle.

When all these symmetries are taken into account, the number of essentially different Sudoku patterns is reduced substantially. In the case of the order 2 Sudoku, it turns out there are actually only two distinct grids! All the rest of the 288 patterns can be generated from these two by applying various symmetry operations (see figure 5.3.). In the order 3 case, the reduction is also dramatic, although it still leaves an abundance of genuinely different solutions. By carefully enumerating the symmetries, Jarvis and

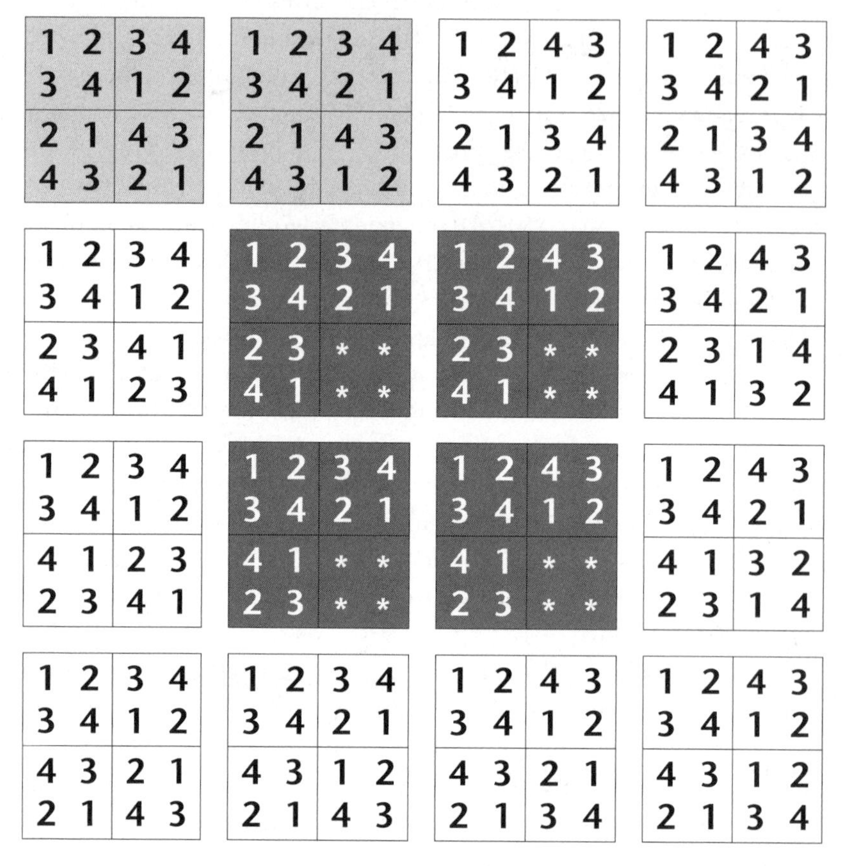

Figure 5.3 Order 2 Sudoku are small enough that all possible configurations can be conveniently enumerated. When the block in the upper left quadrant of the grid is held fixed, the rules of Sudoku allow four variations in the upper right quadrant and four more in the lower left, generating the 16 grids shown here. For 12 of these grids, the lower right quadrant can be completed in just one way; in the remaining four cases *(dark gray grids in center)*, no legal completion of the grid is possible. Thus there are a total of 12 solutions for the given configuration of the upper left quadrant. But that quadrant can actually have any of 24 arrangements of the four numbers, and so the total number of grids is 12×24, or 288. In another sense, there are only two distinct grids. The entire set of 288 solutions can be generated from the two arrangements at the upper left *(light gray grids)*.

Ed Russell came up with the number of unique order 3 Sudoku solutions: 5,472,730,538, or about 5 billion.

Note that this is the number of *solutions*; the number of *puzzles* is vastly greater. For each solution, you can create a puzzle by erasing some subset of the filled-in cells and leaving the rest as givens. Suppose you want a puzzle with 25 givens, leaving the rest of the 81 cells blank. There are about 5×10^{20} ways of choosing 25 items from a set of 81, and each of these subsets yields a potential puzzle. Not all the patterns formed in this way will be valid Sudokus with a unique solution; still, the number of puzzles per solution is enormous.

In the middle of solving a Sudoku, I occasionally have a feeling of déjà vu. The arrangement of numbers in the grid looks familiar, or maybe it's just the pattern of relations between them. Have the publishers been recycling old puzzles? I suppose it's possible, but there's clearly no shortage of fresh puzzles waiting to be solved.

From Gentle to Diabolical

Computer science has an elaborate hierarchy for classifying problems according to difficulty, and it's natural to ask where Sudoku fits into this scheme. Can you write a computer program that will quickly and efficiently solve any Sudoku, or are there puzzles that will cause any program to bog down and run for eons? There are some answers to these questions. Unfortunately, they have almost no bearing on the difficulty of solving the Sudoku you find in today's newspaper, whether you do it by hand or with a computer.

In computer science the great divide is between polynomial and exponential algorithms. Given a problem of size n, an algorithm is considered efficient if its running time is proportional to n or n^2 or even n^{100} (these are all polynomial functions of n). But if the running time grows as 2^n or e^n or n^n (these are exponential functions), the algorithm will be hopelessly slow for large n.

According to Takayuki Yato and Takahiro Seta of the University of Tokyo, Sudoku belongs to a class of problems known as NP-complete. No polynomial algorithm has been found for any of these problems, and most computer scientists consider it unlikely—maybe unthinkable—that fast methods will ever be discovered. One reason for this pessimistic view is that all NP-complete problems are linked together; solve one and you solve them all. The idea that hundreds of seemingly intractable problems will become easy at a single stroke seems too good to be true.

When you lay down your pencil on a completed Sudoku, the thought that you've just dispatched an NP-complete problem may boost your psychological well-being, but don't get carried away; it doesn't mean you have superpowers. The NP-complete label doesn't say anything at all about the difficulty of solving individual Sudoku puzzles. It just states that the effort required will grow exponentially when you turn to puzzles on larger and larger grids.

Some publishers rank their Sudokus on a scale from easy to hard (or from gentle to diabolical). The criteria for these ratings are not stated, and it's a common experience to breeze through a "very hard" puzzle and then get stuck on a "medium." Thus it appears the problem of estimating Sudoku difficulty has not yet been reduced to an exact science.

One easily measured factor that might be expected to influence difficulty is the number of givens. In general, having fewer cells specified at the outset would seem to make for a harder puzzle. At the other extreme, there's no doubt that having a majority of the cells filled in makes a puzzle very easy indeed. In trying to create a Sudoku with fewer givens, the challenge is to ensure that the puzzle has just one solution. After all, a completely blank order 3 grid has more than 5 billion solutions.

What is the minimum number of givens that guarantees uniqueness? On the order 2 grid there are uniquely solvable puzzles with four givens but not, I think, with three. (Finding the arrangements with just four givens is itself a diverting puzzle.) For order 3, the minimum number of givens remained unknown for several years. Gordon Royle of the University of Western Australia collected almost 50,000 examples of uniquely solvable grids with 17 givens, and he found none with fewer than 17. This is strong evidence for the supposition that 17 is the minimum number, but the conjecture was not proved until 2012, when Gary McGuire, Bastian Tugemann, and Gilles Civario of University College Dublin settled the matter. They did it by exhaustive testing of all 5 billion solutions, though they relied on some crucial shortcuts to avoid examining all possible arrangements of 16 clues in each grid. The computation took a year on a cluster of computers with 320 nodes.

The puzzles found in newspapers and other publications seldom have fewer than 20 givens, and the norm is between 25 and 30. Within this range, the correlation between number of givens and difficulty rating is weak. In one book, I found that the gentle puzzles averaged 28.3 givens and the diabolical ones 28.0.

Logic Rules

Many puzzle constructors distinguish between puzzles that can be solved by logic alone and those that require trial and error. If you solve by logic, you never write a number into a cell until you can prove that the number must appear in that position. Trial and error allows for guessing; you fill in a number tentatively, explore the consequences, and if necessary, backtrack, removing your choice and trying another. A logic solver can work with a pen; a backtracker needs a pencil and an eraser.

For the logic-only strategy to work, a puzzle must have a quality of progressivism; at every stage in the solution, there must be at least one cell whose value can be determined unambiguously. Filling in that value must then uncover at least one other fully determined value, and so on. The backtracking protocol dispenses with progressivism. When you reach a state where no choice is forced upon you—where every vacant cell has at least two candidates—you choose a path arbitrarily.

The distinction between logic and backtracking seems like a promising criterion for rating the difficulty of puzzles, but on a closer look, it's not clear the distinction even exists. Is there a subset of Sudoku puzzles that can be solved by backtracking but not by logic alone? The issue can be made clearer by asking the question another way: Are there puzzles that have a unique solution, and yet at some intermediate stage reach an impasse, where no cell has a value that can be deduced unambiguously? Not, I think, unless we impose artificial restrictions on the rules allowed in making logical deductions.

Backtracking itself can be viewed as a logical operation; it supplies a proof by contradiction. If you make a speculative entry in one cell and, as a consequence, eventually find that some other cell has no legal entry, then you have discovered a logical relation between those two cells. The chain of implication could be very intricate, but the logical relation is no different in kind from the simple rule that says two cells in the same row can't have the same value. (David Eppstein has formulated some extremely subtle Sudoku rules, which capture the kind of information gleaned from a backtracking analysis, yet work in a forward-looking, nonspeculative mode.)

A Satisfied Mind

From a computational point of view, Sudoku is a constraint satisfaction problem. The constraints are the rules forbidding two cells in the same neighborhood to hold the same number; a solution is an assignment of numbers to cells that satisfies all the constraints simultaneously. One encoding of the problem for the order 3 Sudoku imposes just 27 constraints, one for each row, each column, and each block. These constraints take the form of an all-different rule; all the cells in the neighborhood must have different values. The same logic can be expressed by simpler not-equal rules on pairs of individual cells. This approach has the advantage that programming languages offer a built-in not-equal operator. On the other hand, the total number of rules is much larger: 810.

It's interesting to observe how differently one approaches the Sudoku puzzle when solving it by computer rather than by hand. A human solver may well decide that logic is all you need, but backtracking is the more appealing option for a program. For one thing, backtracking will always find the answer, if there is one. It can even do the right thing if there are multiple solutions or no solution. To make similar claims for a logic-only program, you would have to prove you had included every rule of inference that might possibly be needed.

Backtracking is also the simpler approach, in the sense that it relies on one big rule rather than many little ones. At each stage you choose a value for some cell and check to see if this new entry is consistent with the rest of the grid. If you detect a conflict, you have to undo the choice and try another. If you have exhausted all the candidates for a given cell without finding a value that can be placed there, then you must have taken a wrong turn earlier, and you need to backtrack further. This method of puzzle solving is not a clever algorithm; it amounts to a depth-first search of the tree of all possible solutions—a tree that could have 9^{81} leaves. There is no question that we are deep in the territory of exponential algorithms here. And yet, in practice, solving Sudoku by backtracking is embarrassingly easy.

There are many strategies for speeding up the backtracking search process, mostly focused on making a shrewd choice of which branch of the tree to try next. But such optimizations are hardly needed. On an order 3 Sudoku grid, even a rudimentary backtracking search converges on the solution in a few dozen steps. The fastest publicly available programs solve hard Sudokus in about 1 millisecond. Evidently, competing against a computer in Sudoku is never going to be much fun.

Does that ruin the puzzle for the rest of us? In moments of frustration, when I'm struggling with a recalcitrant diabolical, the thought that the machine across the room could instantly sweep away all my cobwebs of logic is indeed dispiriting. I begin to wonder whether this cross-correlation of columns, rows, and blocks is a fit task for the human mind. But when I do make a breakthrough, I take more pleasure in my success than the computer would.

6

Crinkly Curves

In 1877 the German mathematician Georg Cantor made a shocking discovery. He found that a two-dimensional surface contains no more points than a one-dimensional line. Cantor compared the set of all points forming the area of a square with the set of points along one of the line segments on the perimeter of the square. He showed that the two sets are the same size. Intuition rebels against this notion. Inside a square you could draw infinitely many parallel line segments side by side. Surely an area with room for such an infinite array of lines must include more points than a single line—but it doesn't. Cantor himself was incredulous: "I see it, but I don't believe it," he wrote.

Yet the fact was inescapable. Cantor defined a one-to-one correspondence between the points of the square and the points of the line segment. Every point in the square was associated with a single point in the segment; every point in the segment was matched with a unique point in the square. No points were left over or used twice. It was like pairing up mittens: If you come out even at the end, you must have started with equal numbers of lefts and rights.

Geometrically, Cantor's one-to-one mapping is a scrambled affair. Neighboring points on the line scatter to widely separated destinations in the square. The question soon arose, Is there a *continuous* mapping between a line and a surface? In other words, can one trace a path through a square without ever lifting the pencil from the paper and touch every point at least once? It took a decade to find the first such curve. Then dozens more were invented, as well as curves that fill up a three-dimensional volume or even a region of some higher-dimensional space. The very concept of dimension was undermined.

Circa 1900 these space-filling curves were viewed as mysterious aberrations, signaling how far mathematics had strayed from the world of everyday experience. The mystery has never entirely faded away, but the curves have grown more familiar. They are playthings of programmers now, nicely adapted to illustrating certain algorithmic techniques (especially recursion). More surprising, the curves have turned out to have practical applications. They serve to encode geographic information; they have a role in image processing; they help allocate resources in large computing tasks. And they tickle the eye of those with a taste for intricate geometric patterns.

How to Fill Up Space

It's easy to sketch a curve that completely fills the interior of a square. The finished product looks like this:

How uninformative! It's not enough to know that every point is covered by the passage of the curve; we want to see how the curve is constructed and what route it follows through the square.

If you were designing such a route, you might start out with the kind of path that's good for mowing a lawn:

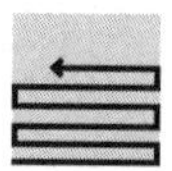 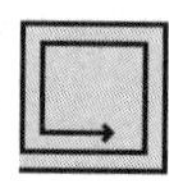

But there's a problem with these zigzags and spirals. A mathematical lawn mower cuts a vanishingly narrow swath, and so you have to keep reducing the space between successive passes. Unfortunately, the limiting pattern when the spacing goes to zero is not a filled area; it is a path that forever retraces the same line along one edge of the square or around its perimeter, leaving the interior blank.

The first successful recipe for a space-filling curve was formulated in 1890 by Giuseppe Peano, an Italian mathematician also noted for his axioms of arithmetic. Peano did not provide a diagram or even an explicit description of what his curve might look like; he merely defined a pair of mathematical functions that give x and y coordinates inside a square for each position t along a line segment.

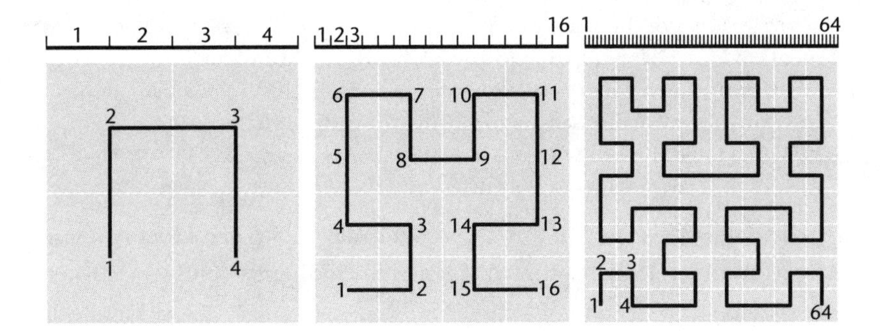

Figure 6.1 A space-filling curve evolves through successive stages of refinement as it grows to cover the area of a square. This illustration is a redrawing of the first published diagram of such a curve; the original appeared in an 1891 paper by David Hilbert. The idea behind the construction is to divide a line segment into four intervals and divide a square into four quadrants, then establish a correspondence between the points of corresponding intervals and quadrants. The process continues with further recursive subdivisions.

Soon David Hilbert, a leading light of German mathematics in that era, devised a simplified version of Peano's curve and discussed its geometry. Figure 6.1 is a redrawing of a diagram from Hilbert's 1891 paper, showing the first three stages in the construction of the curve.

Programming by Procrastination

Figure 6.2 shows a later stage in the evolution of the Hilbert curve, when it has become convoluted enough that one might begin to believe it will eventually reach all points in the square. The curve was drawn by a computer program written in a recursive style that I call programming by procrastination. The philosophy behind the approach is this: Plotting all those twisty turns looks like a tedious job, so why not put it off as long as we can? Maybe we'll never have to face it.

Let us eavesdrop on a computer program named Hilbert as it mumbles to itself while trying to solve this problem:

> Hmm. I'm supposed to draw a curve that fills a square. I don't know how to do that, but maybe I can cut the problem down to size. Suppose I had a subroutine that would fill a smaller square, say one-fourth as large. I could invoke that procedure on each quadrant of the main square, getting back four

separate pieces of the space-filling curve. Then, if I just draw three line segments to link the four pieces into one long curve, I'll be finished!

Of course I don't actually have a subroutine for filling in a quadrant. But a quadrant of a square is itself a square. There's a program named Hilbert that's supposed to be able to draw a space-filling curve in any square. I'll just hand each of the quadrants off to Hilbert.

The strategy described in this monologue may sound like a totally pointless exercise. The Hilbert program keeps subdividing the problem but has

Figure 6.2 After seven stages of elaboration the Hilbert curve meanders through 4^7 = 16,384 subdivisions of the square. The curve is continuous in the sense that it has no gaps or jumps, but it is not smooth: All the right angles are points where the curve has no tangent (or, in terms of calculus, no derivative). Continuing the subdivision process leads to a limiting case where the curve fills the entire square, showing that a two-dimensional square has no more points than a one-dimensional line segment.

no plan for ever actually solving it. However, this is one of those rare and wonderful occasions when procrastination pays off, and the homework assignment you lazily set aside last night is miraculously finished when you get up in the morning.

Consider the sizes of the successive subsquares in Hilbert's divide-and-conquer process. At each stage, the side length of the square is halved, and the area is reduced to one-fourth. The limiting case, if the process goes on indefinitely, is a square of zero side length and zero area. So here's the procrastinator's miracle: Tracing a curve that touches all the points inside a size zero square is easy because such a square is in fact a single point. Just draw it!

Practical-minded readers will object that a program running in a finite machine for a finite time will not actually reach the limiting case of squares that shrink away to zero size. I concede the point. If the recursion is halted while the squares still contain multiple points, one of those points must be chosen as a representative; the center of the square is a likely candidate. In plotting the convoluted maze of figure 6.2, I stopped the program after seven levels of recursion, when the squares were small but certainly larger than a single point. The wiggly black line connects the centers of $4^7 = 16{,}384$ squares. Only in the mind's eye will we ever see a truly infinite space-filling curve, but a finite drawing like this one is at least a guide to the imagination.

I have glossed over another important aspect of this algorithm. If the curve is to be continuous, with no abrupt jumps, then all the squares have to be arranged so that one segment of the curve ends where the next segment begins. Matching up the end points in this way requires rotating and reflecting some of the subsquares. (For an animated illustration of these transformations, see http://bit-player.org/extras/hilbert.)

Grammar and Arithmetic

The procrastinator's algorithm is certainly not the only way to draw a space-filling curve. Another method exploits the self-similarity of the pattern—the presence of repeated motifs that appear in each successive stage of the construction. In the Hilbert curve the basic motif is a U-shaped path with four possible orientations. In going from one stage of refinement to the next, each U orientation is replaced by a specific sequence of four smaller U curves, along with line segments that link them together,

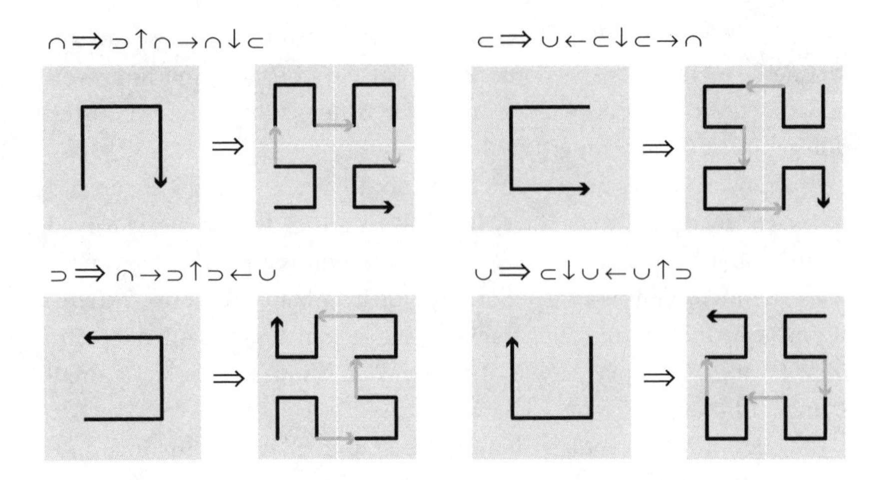

Figure 6.3 Substitution rules generate the Hilbert curve by replacing a U-shaped motif in any of four orientations with sequences of four rotated and reflected copies of the same motif. The set of rules taken together constitute a grammar.

as shown in figure 6.3. The substitution rules form a grammar that generates geometric figures in the same way that a linguistic grammar generates phrases and sentences.

The output of the grammatical process is a sequence of symbols. An easy way to turn it into a drawing is to interpret the symbols as commands in the language of "turtle graphics." The turtle is a conceptual drawing instrument that crawls over the plane in response to simple instructions to move forward, turn left, or turn right. The turtle's trail across the surface becomes the curve to be drawn.

When Peano and Hilbert were writing about the first space-filling curves, they did not explain them in terms of grammatical rules or turtle graphics. Instead their approach was numerical, assigning a number in the interval [0, 1] to every point on a line segment and also to every point in a square. For the Hilbert curve, it's convenient to do this arithmetic in base 4, or quaternary, working with the digits 0, 1, 2, 3. In a quaternary fraction such as 0.213, each successive digit specifies a quadrant or subquadrant of the square, as outlined in figure 6.4.

What about other space-filling curves? Peano's curve is conceptually similar to Hilbert's but divides the square into nine regions instead of four. Another famous example was invented in 1912 by the Polish mathematician Wacław

Sierpiński; it partitions the square along its diagonals, forming triangles that are then further subdivided. A more recent invention is the flowsnake curve, devised in the 1970s by Bill Gosper (see figure 6.5.).

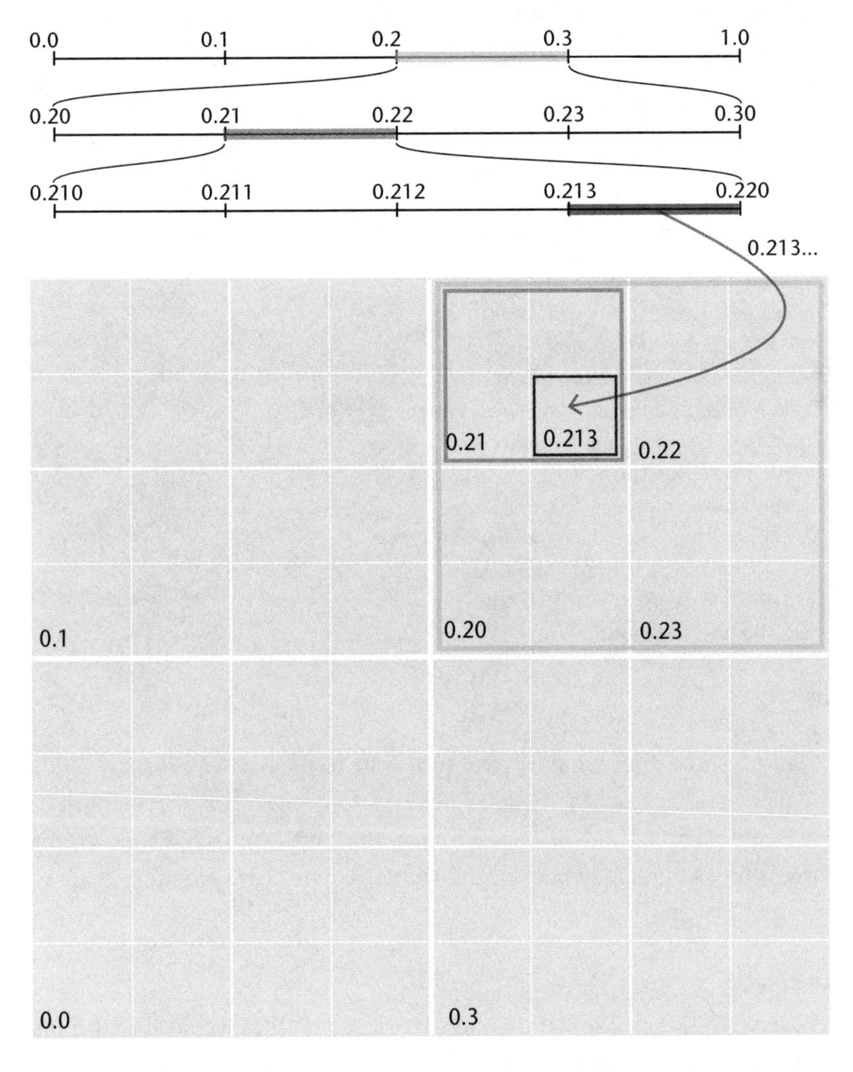

Figure 6.4 Base-4 encoding of the Hilbert curve shows how fourfold divisions of the unit interval [0, 1] are mapped onto quadrants of the square. For example, any base-4 number beginning 0.2... correponds to a point somewhere in the upper right quadrant, outlined in light gray. All numbers beginning 0.21... go into the square outlined in medium gray, and those in 0.213... lie in the small square outlined in dark gray.

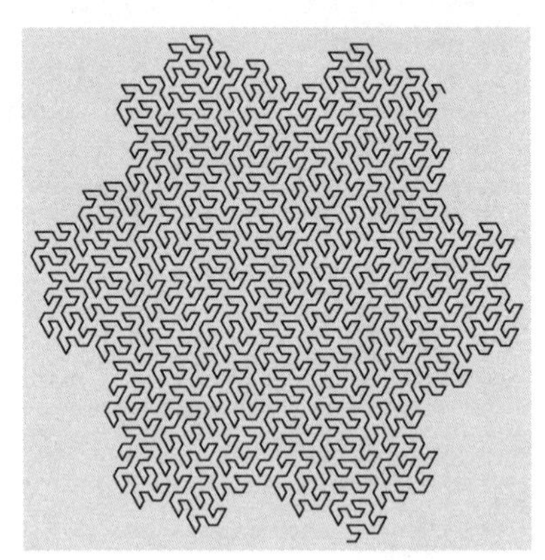

Figure 6.5 The first space-filling curve *(above left)* was described in 1890 by the Italian mathematician Giuseppe Peano; the construction divides a square into nine smaller squares. A curve based on a triangular dissection *(above right)* was introduced in 1912 by the Polish mathematician Wacław Sierpiński. The flowsnake curve *(right)*, devised by the American mathematician Bill Gosper in the 1970s, fills a ragged-edged hexagonal area.

Filling three-dimensional space turns out to be even easier than filling the plane—or at least there are more ways to do it. Herman Haverkort of the Eindhoven Institute of Technology in the Netherlands has counted the three-dimensional analogues of the Hilbert curve; there are more than 10 million of them.

All Elbows

In everyday speech the word *curve* suggests something smooth and fluid, without sharp corners, such as a parabola or a circle. The Hilbert curve is anything but smooth. All finite versions of the curve consist of 90-degree bends connected by straight segments. In the infinite case, where the contruction process is not halted after any finite number of iterations, the straight segments dwindle away to zero length, leaving nothing but sharp corners. The curve is all elbows. In 1900 the American mathematician Eliakim Hastings Moore came up with the term *crinkly curves* for such objects.

Even though the complete path of an infinite space-filling curve cannot be drawn on paper, it is still a perfectly well-defined object. You can

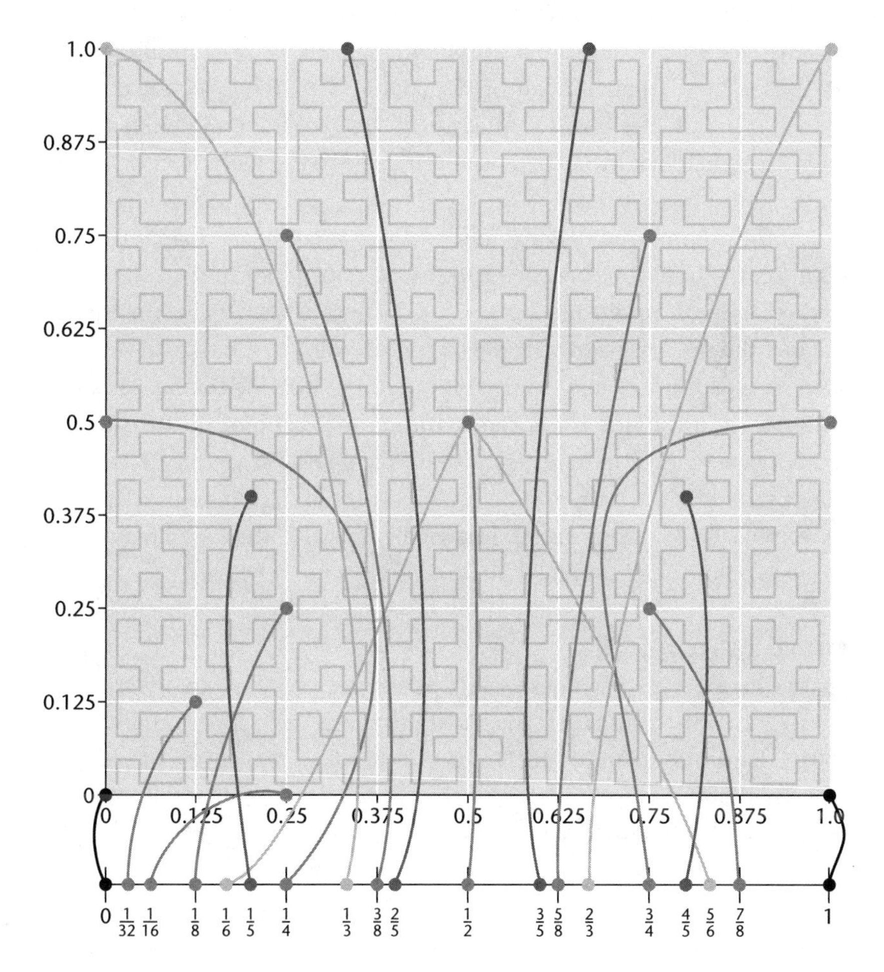

Figure 6.6 Positions of points along the infinitely crinkled course of the Hilbert curve can be calculated exactly, even though the infinite curve itself cannot be drawn. Here 25 selected points in the interval $[0,1]$ are mapped to coordinates in the unit square, $[0,1]^2$. Although a finite approximation to the Hilbert curve is shown in the background, the positions within the square are those along the completed infinite curve. The inverse mapping is not unique; some points in the square map back to multiple points in the interval.

calculate the location along the curve of any specific point. The result is exact if the input is exact. Figure 6.6 shows some landmark points along the two-dimensional Hilbert curve and their corresponding positions on the one-dimensional line segment.

The algorithm for this calculation directly implements the definition of the curve as a mapping from a one-dimensional line segment to a two-dimensional square. The input to the function is a number in the interval [0, 1], and the output is a pair of *x, y* coordinates. The inverse mapping—from *x, y* coordinates back to the segment [0, 1]—is more troublesome. The problem is that a point in the square can be linked to more than one point on the line.

Cantor's dimension-defying function was a *one-to-one* mapping. Each point on the line was associated with exactly one point in the square, and vice versa. But Cantor's mapping was not continuous; adjacent points on the line did not necessarily map to adjacent points in the square. In contrast, the space-filling curves are continuous but not one-to-one. Although each point on the line is associated with a unique point in the square, a point in the square can map back to multiple points on the line. A conspicuous example is the center of the square, with the coordinates $x = \frac{1}{2}$, $y = \frac{1}{2}$. Three separate locations on the line segment ($\frac{1}{6}$, $\frac{1}{2}$, and $\frac{5}{6}$) all connect to this one point in the square.

Math on Wheels

Space-filling curves have been called monsters, but they are useful monsters. One of their most remarkable applications was reported in 1983 by John J. Bartholdi III and his colleagues at the Georgia Institute of Technology. Their aim was to find efficient routes for drivers delivering Meals on Wheels to elderly clients scattered around the city of Atlanta. Finding the best possible delivery sequence would be a challenging task even with a powerful computer. Meals on Wheels didn't need the solution to be strictly optimal, but they needed to plan and revise routes quickly, and they had to do it with no computing hardware at all. Bartholdi and his coworkers came up with a scheme that used a map, a few pages of printed tables, and two Rolodex files.

Planning a route started with Rolodex cards listing the delivery addresses. The manager looked up the map coordinates of each address, then looked up those coordinates in a table, which supplied an index number to write

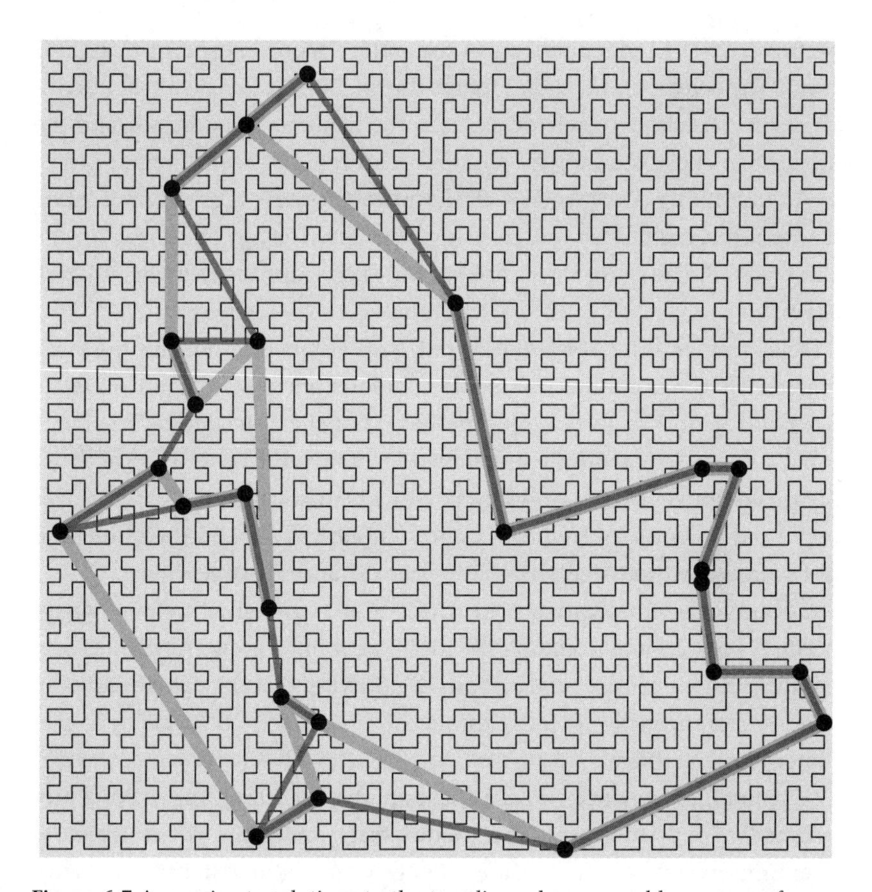

Figure 6.7 Approximate solutions to the traveling salesman problem emerge from a simple algorithm based on space-filling curves. Here 25 cities *(black dots)* are randomly distributed within a square. The traveling salesman problem calls for the shortest tour that passes through all the cities and returns to the starting point. Listing the cities in the order they are visited by a space-filling curve yields a path of length 274 *(thick, light gray line)*; the optimal tour *(thin, dark gray line)* is about 13 percent better, with a length of 239. The space-filling curve used in this example was invented by E. H. Moore in 1900; it is related to the Hilbert curve but forms a closed circuit. The unit of distance for measuring tours is the step size of the Moore curve. The optimal tour was computed with the Concorde TSP Solver (http://www.math.uwaterloo.ca/tsp/concorde/).

on the Rolodex card. Sorting the cards by index number yielded the delivery sequence.

Behind the scenes in this procedure was a space-filling curve (specifically, a finite approximation to a Sierpiński curve) that had been superimposed on the map. The index numbers in the tables encoded positions along this curve. The delivery route didn't follow the Sierpiński curve, with all its crinkly turns. The curve merely determined the sequence of addresses, and the driver then chose the shortest point-to-point route between them.

A space-filling curve works well in this role because it preserves locality. If two points are nearby on the plane, they are likely to be nearby on the curve as well. The route makes no wasteful excursions across town and back again.

The Meals on Wheels scheduling task is an instance of the traveling salesman problem, a notorious stumper in computer science (see figure 6.7). The Bartholdi algorithm gives a solution that is not guaranteed to be best but is usually good. For randomly distributed locations, the tours average about 25 percent longer than the optimum. Other heuristic methods can beat this performance, but they are much more complicated. The Bartholdi method finds a route without even computing the distances between sites.

Locality is a helpful property in other contexts as well. Sometimes what's needed is not a route from one site to the next but a grouping of sites into clusters. In two or more dimensions, identifying clusters can be difficult; threading a space-filling curve through the data set reduces it to a one-dimensional problem.

The graphic arts have enlisted the help of space-filling curves for a process known as halftoning, which allows black-and-white devices (such as laser printers) to reproduce shades of gray. Conventional halftoning methods rely on arrays of dots that vary in size to represent lighter and darker regions. Both random and regular arrays tend to blur fine features and sharp lines in an image. A halftone pattern that groups the dots along the path of a Hilbert or Peano curve can provide smooth tonal gradients while preserving crisp details.

Another application comes from a quite different realm: the multiplication of matrices (a critical step in large-scale computations). Accessing matrix elements by rows and columns requires the same values to be read from memory multiple times. In 2006 Michael Bader and Christoph Zenger of the Technical University of Munich showed that clustering the data with a space-filling curve reduces memory traffic.

Bader is also the author of an excellent 2013 book that discusses space-filling curves from a computational point of view. An earlier volume by Hans Sagan is more mathematical.

Given that people have found such a surprising variety of uses for these curious curves, I can't help wondering whether nature has also put them to work. Other kinds of patterns are everywhere in the natural world: stripes, spots, spirals, and many kinds of branching structures. But I can't recall seeing a Peano curve on the landscape. The closest I can come are certain trace fossils (preserved furrows and burrows of organisms on the sea floor) and perhaps the ridges and grooves on the surface of the human cerebrum.

Cantor's Conundrums

Applications of space-filling curves are necessarily built on finite examples—paths one can draw with a pencil or a computer. But in pure mathematics the focus is on the infinite case, where a line gets so incredibly crinkly that it suddenly becomes a plane.

Cantor's work on infinite sets was controversial and divisive in his own time. Leopold Kronecker, who had been one of Cantor's professors in Berlin, later called him "a corrupter of youth" and tried to block publication of the paper on dimension. But Cantor had ardent defenders, too. Hilbert wrote in 1926, "No one shall expel us from the paradise that Cantor has created." And Hilbert was right. No one has been evicted. (A few have left of their own volition.)

Cantor's discoveries eventually led to clearer thinking about the nature of continuity and smoothness, concepts at the root of calculus and analysis. The related development of space-filling curves called for a deeper look at the idea of dimension. From the time of Descartes, it was assumed that in d-dimensional space it takes d coordinates to state the location of a point. The Peano and Hilbert curves overturned this principle. A single number can define position on a line, on a plane, in a solid, or even in those 11-dimensional spaces so fashionable in high-energy physics.

At about the same time that Cantor, Peano, and Hilbert were creating their crinkly curves, the English schoolmaster Edwin Abbott was writing his fable *Flatland*, about two-dimensional creatures that dream of popping out of the plane to see the world in 3-D. The Flatlanders might be encouraged to learn that mere one-dimensional worms can break through to higher spaces just by wiggling wildly enough.

7

Wagering with Zeno

Vacationing in Italy, you wander into the coastal village of Velia, a few hours south of Naples. On the edge of town you notice an archaeological dig. When you go to have a look at the ruins, you learn that the place now called Velia was once the Greek settlement of Elea, home to the philosopher Parmenides and his disciple Zeno. You stroll through the excavated baths and trace the city walls, then climb a steep, cobbled roadway to an arch called the Porta Rosa (see figure 7.1). Perhaps Zeno formulated his famous paradoxes while pacing these same stones 900,000 days ago. Was there something special about the terrain that led him to imagine arrows frozen in flight and runners who go halfway, then half the remaining half, but never get to the finish line?

That night, Zeno visits you in a dream. He brings along a sack of ancient coins, which come in denominations of 1, ½, ¼, ⅛, 1⁄16, and so on. Evidently the Eleatic currency had no smallest unit; for every coin of value $1/2^n$, there is another of value $1/2^{n+1}$. Zeno's bag holds exactly one coin of each denomination.

He teaches you a gambling game. First the coin of value 1 is set aside; it belongs to neither of you but will be flipped to decide the outcome of each round of play. Now the remaining coins are divided in such a way that each of you has a total initial stake of exactly ½. The distinctively Eleatic part of the game is the rule for setting the amount of the wager. Before each coin toss, you and Zeno each count your current holdings, and the bet is one-half of the lesser of these two amounts. Thus the first wager is ¼. Suppose you win that toss. After the bet is paid, you have ¾, and Zeno's fortune is reduced to ¼; the amount of the next bet is therefore ⅛. Say Zeno wins this time; then the score stands at ⅝ for you and ⅜ for him, and the next amount at stake is 3⁄16. If Zeno wins again, he takes the lead, 9⁄16 to 7⁄16.

In the morning you wake up wondering about this curious game. What is the likely outcome if you continue playing indefinitely? Is one player sure to win eventually, or could the lead be traded back and forth forever?

Can't Win, Can't Lose, Can't Tie

A few properties of the Zeno game are easy to state. For example, the betting process appears to be fair (assuming that the coin being flipped is unbiased). Each player has the same odds of winning or losing each round, and the amount at risk is the same.

Another way of saying that the game is fair is that the expectation value for each player is ½. If you play many independent games, you should

Figure 7.1 A stone gate called the Porta Rosa once connected two quarters of the ancient city of Elea, near what is now the town of Velia in southern Italy. Elea was the home of Zeno, the philosopher famous for his paradoxes of multitude and motion. The seemingly redundant "eyebrow" directly above the main arch of the Porta Rosa may have had a structural role, relieving stresses in the stonework. But followers of Zeno's thinking could tell a different story. If Zeno had designed the gate, he would have placed the primary arch halfway between the ground and the top of the wall, then a second arch halfway between the first arch and the top, a third arch halfway again, and so on. Perhaps the repairs near the top of the wall testify to the impracticality of this scheme?

come out roughly even in the end. But an expectation value of ½ does *not* mean you should expect to go home with half the money at the end of a single game. Indeed, after the first coin toss, the game cannot possibly end in a tie.

But you can never go broke, either—at least not in a finite number of plays. However small your remaining wealth, the wagering rule says you can't risk more than half of it. Of course, the same reasoning protects your opponent as well; if you can't lose everything, neither can you win it all.

Here's another observation: In the game-within-a-dream, all the numbers mentioned have a distinctive appearance. They are fractions whose denominator is a power of 2. In other words, they are numbers of the form $m/2^n$, called dyadic rationals. Is this predilection for halves, fourths, eighths, sixteenths, and so on, a peculiarity of that one example, or does the pattern carry over to all Zeno games?

The answer comes from an inductive argument. Suppose at some stage of the game your score is a dyadic rational, x, and is less than or equal to ½. Then the amount at stake in the next round of wagering is $x/2$, so that your new tally will be either $x - x/2$ or $x + x/2$. But $x - x/2$ is simply $x/2$, and $x + x/2$ is $3x/2$; both these numbers are dyadic rationals. A similar (but messier) argument establishes the same result for values of x greater than ½. Thus if your score is ever a dyadic rational, it will remain one for the rest of the game. But the starting value, ½, is itself a dyadic rational, and so the only numbers that can ever arise in the game are fractions of the form $m/2^n$.

This line of argument actually yields a slightly stronger result. For a score $x < 1/2$, the net effect of the gambling transaction is to multiply x by either ½ or $3/2$. In either case, the denominator is doubled; as the game proceeds, the denominator increases monotonically. An important consequence is that the entire numerical process is *nonrecurrent.* In the course of a game you'll never see the same number twice. This is one reason the game can't end in a tie. After the first flip of the coin, the score can never find its way back to ½.

Walking with Zeno

The evolution of a Zeno wagering game corresponds to a special kind of random walk. A player's gains and losses are represented by the movement of a walker along the interval between 0 and 1. The walker starts at the position $x = 1/2$. Each flip of the coin determines whether the next step is to the left (toward 0) or the right (toward 1). The length of the step is half the distance

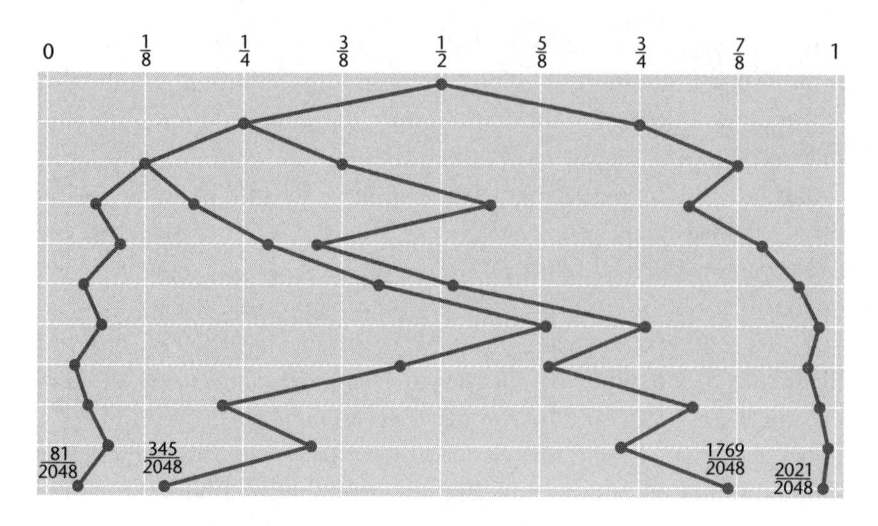

Figure 7.2 In Zeno's wagering game, two players start with a stake of ½ each and decide a series of bets by flipping a fair coin. The size of each bet is half the current wealth of the poorer player. The game is equivalent to a random walk on the interval from 0 to 1, taking steps whose size is equal to half the distance to the nearer boundary. The trajectories of four such walks are shown here.

to whichever of these boundaries is nearer. In other words, the step length is $\frac{1}{2}\min(x, 1-x)$.

Figure 7.2 shows a few trajectories constructed according to these rules. One feature of note is an apparent tendency for paths to flee the middle of the interval and linger near the edges. It's not hard to understand this behavior, at least in a qualitative way. Whenever the walker is near the center, it is moving with higher velocity (that is, taking larger steps per unit time), and so it doesn't stay long in this neighborhood. Out at the periphery, the walker moves very slowly, and so it takes a long time to escape. It's as if the walker were moving over a landscape that's smoothly paved in the middle but becomes a sticky mire near the edges.

A plausible hypothesis suggests that a typical random walk will spend more and more time near the end points of the interval as the walk proceeds, coming arbitrarily close to 0 and 1. To test this idea you might follow a walk for many thousands of steps, but that process is computationally challenging. If you represent the walker's position by means of a floating-point number, the program will usually report that the walker has reached either 0.0 or 1.0 after just a few hundred steps. This outcome would surprise Zeno! The problem is that floating-point formats have only finite precision, and very small values are rounded to zero.

A remedy for the round-off problem is exact rational arithmetic, but this becomes cumbersome. Here is the unwieldy numerical value of a game score after 150 steps:

$$\frac{285449538541182705265342404106190451084008266l}{285449538541191976211657193889899027276549324 8}$$

The numerator and denominator both have 45 digits, and they differ by less than one part in a trillion.

Zeno's Favorite Numbers

An alternative to tracing a few very long games is to gather statistics on the outcome of many shorter games. Figure 7.3 gives the observed frequencies of various outcomes for games of length 1 through 6, based on samples of several thousand trials.

Games of length 1 (a single coin toss) can have only two possible outcomes, namely, $\frac{1}{4}$ and $\frac{3}{4}$, and these events are equally likely. Two-round games must end with a value of $\frac{1}{8}$, $\frac{3}{8}$, $\frac{5}{8}$, or $\frac{7}{8}$, and again all four choices have the same probability.

Things get interesting with games of three or more rounds. After the third coin toss, the score of the gambler (or the position of the random walker) must be a fraction that, when expressed in lowest terms, has a denominator of 16. There are eight such fractions, but only six of them ever turn up as results of Zeno games; $\frac{5}{16}$ and $\frac{11}{16}$ are simply not observed. Among the six values that *do* occur, two of them ($\frac{3}{16}$ and $\frac{13}{16}$) are twice as common as the others.

Going on to four-round games, the pattern gets more peculiar. In this case all game values must be fractions with a denominator of 32. Of the 16 possibilities, only 10 are actually observed, and a few of these are two or three times more frequent than others. The likeliest game outcomes are $\frac{3}{32}$ and $\frac{9}{32}$ (along with the symmetrically related values $\frac{29}{32}$ and $\frac{23}{32}$, which are equal to $1-\frac{3}{32}$ and $1-\frac{9}{32}$). The differences in frequency are much too large to be an effect of statistical noise.

As the number of wagering rounds increases further, the patterns become even more pronounced. Wide gaps in the frequency distribution turn the graph into a snaggle-tooth smile. And certain numbers are dramatically more

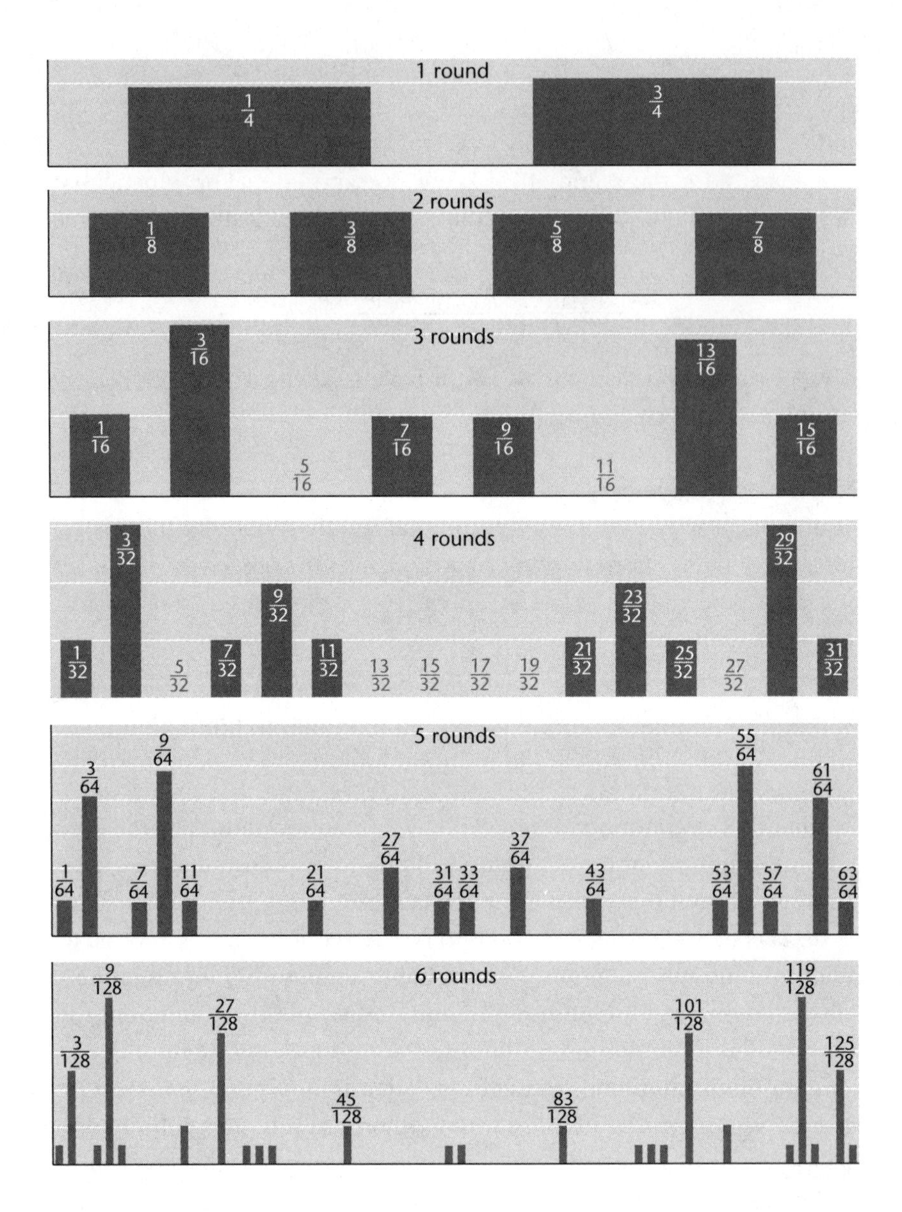

Figure 7.3 The statistics of the Zeno game show that not all outcomes are equally likely, except in the very shortest games. The charts record the observed frequency of all possible end states for games that last from one round (that is, a single coin toss) to six rounds. The only numbers that can appear in the game are dyadic rationals—fractions of the form $m/2^n$—but only a subset of these numbers are actually observed. Among the numbers present, some are much more popular than others; especially common are those with a power of 3 in the numerator (3, 9, 27...).

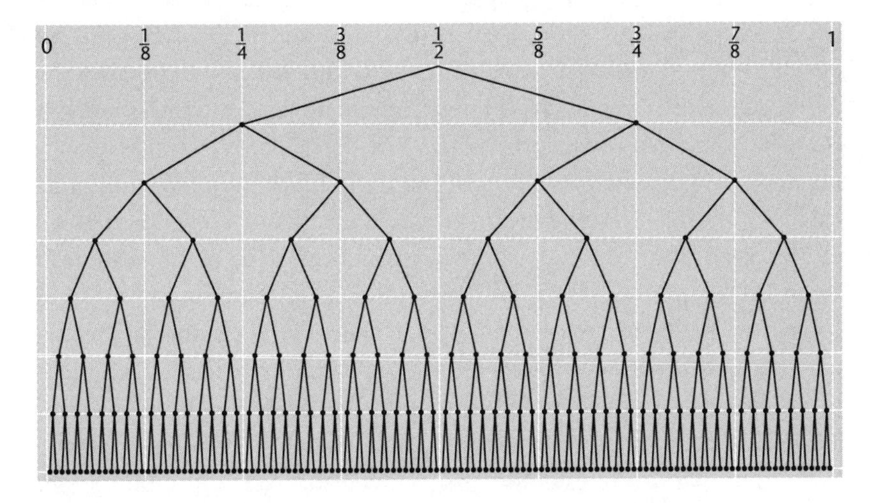

Figure 7.4 A symmetrical binary tree represents all possible trajectories of a game similar to but simpler than the Zeno game. In this variant, the size of the wager—or the length of a step in a random walk—is cut in half after every round. All the dyadic rationals are present in the tree, and at each level they all have the same probability. Paths within the tree never cross or touch. The structure is called a binary tree because each node has two children; each child has just one parent node.

popular than the rest. For games of length 6, only 24 of 64 possible outcomes are observed, and much of the probability is concentrated in just three values (and their symmetrical counterparts). The three favored fractions are $9/128$, $27/128$, and $3/128$. Why does the Zeno process favor these particular numbers? The powers of 2 in the denominator have already been explained, but why do the most common game outcomes all have powers of 3 in the numerator? It can't be an accident.

Climbing Zeno's Tree

In an effort to puzzle out these patterns, I tried constructing the tree of all possible outcomes for games of a given depth. In other words, I listed the two available moves from the starting state $x = 1/2$; then for each of these positions, I wrote down the two possible outcomes of another coin flip, and so on. (Eventually I wrote a program to do the calculations and draw the tree.)

Before plunging into the intricacies of the Zeno tree, it's worth pausing to look at a simpler model—a random walk in which the step size is halved with every step (see figure 7.4). Starting at $x = 1/2$, the walker goes either left or right a distance of 1⁄4, then 1⁄8, then 1⁄16, and so on. Tabulating all possible walks of this kind yields a binary tree that fans out to reach all the dyadic rationals. The branches bifurcate symmetrically, and each branch is completely isolated from all the others. The descendants of two neighboring nodes will come arbitrarily close but never touch. In a wagering game based on this rule, winning the first round is enough to put you in the lead forever. Even if you lose every subsequent bet, your share of the wealth can never fall below 1⁄2.

The Zeno game tree (see figure 7.5) starts out exactly like this tidy binary tree—the first three levels are identical—but then things start to get weird. In the lower levels of the tree there's a lot of disorder, with various gaps between adjacent nodes, and many crossing branches. Even more remarkable, many paths bifurcate and then immediately come together again.

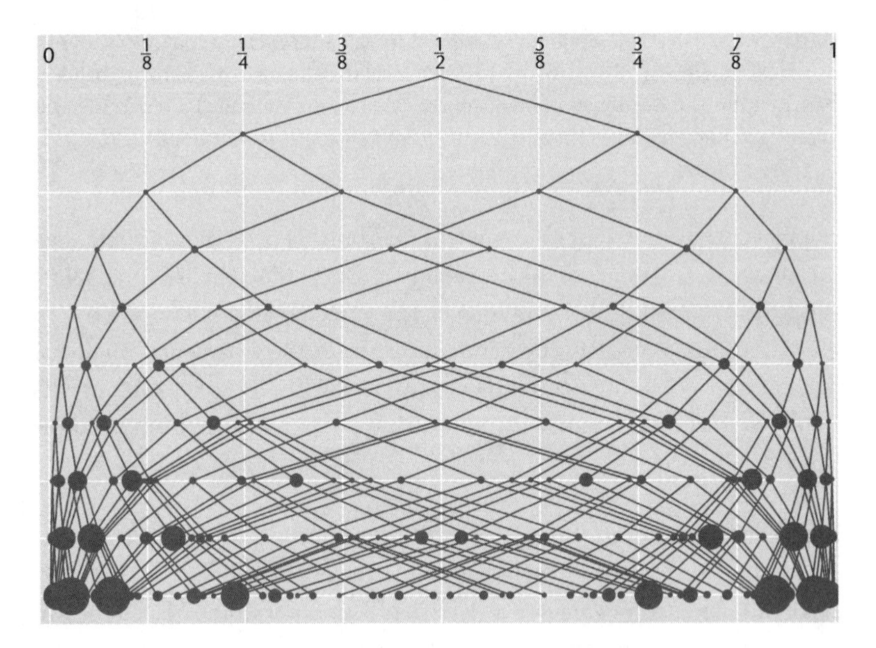

Figure 7.5 All possible trajectories of the Zeno game form a binary tree, but the structure is highly irregular. Many branches cross, and there are also mergers, where a single node is a child of two sibling parents. Because of such rejoinings, many nodes can be reached by more than one path from the tree's root. The probability that a game will end at a particular node is proportional to the number of pathways reaching it, indicated here by the size of the dots marking the nodes.

(Strictly speaking, a structure with such merging branches isn't a tree at all, but it will do no harm to continue using the term.)

One can understand a lot of what's going on in the Zeno tree by looking at two specific fragments. First is the loop formed below the node at $\frac{1}{4}$. One path descending from this node goes left to $\frac{1}{8}$ and then right to $\frac{3}{16}$; another path goes right to $\frac{3}{8}$ and then left to meet up with its sibling at $\frac{3}{16}$. The fact that these two routes converge on precisely the same point is not some miraculous numerical coincidence; it's just a matter of working out the arithmetic:

$$x - \frac{x}{2} + \frac{x - \frac{x}{2}}{2} = x + \frac{x}{2} - \frac{x + \frac{x}{2}}{2}.$$

When the dust settles, this identity comes down to the proposition that a half of three halves is three quarters, and likewise three halves of a half is three quarters.

The second interesting spot in the tree is right next door—the pair of paths descending from the node at $\frac{3}{8}$. The branch to the left, as we've already seen, goes to $\frac{3}{16}$, and then the next turn right on this path lands at $\frac{9}{32}$. The alternative route heads right from $\frac{3}{8}$ to $\frac{9}{16}$; however, on the next leftward bend this path fails to rejoin its partner at $\frac{9}{32}$. Instead the branch stops short at $\frac{11}{32}$. The reason is that this path crosses the midline of the tree at $x = \frac{1}{2}$, and for points to the right of this line, distance is measured not from 0 but from 1. The two branches fail to meet because the equation no longer holds:

$$x - \frac{x}{2} + \frac{x - \frac{x}{2}}{2} \neq x + \frac{x}{2} - \frac{1 - \left(x + \frac{x}{2}\right)}{2}.$$

This mechanism effectively divides the tree into three vertical zones. For all points to the left of $\frac{1}{3}$ and to the right of $\frac{2}{3}$, a node's two children are both on the same side of the midline and thus are governed by the same rule for calculating step lengths. As a result, these zones form a fairly regular lattice-like structure made up of diamond-shaped panes that grow narrower toward the periphery. In the middle zone, by contrast, all nodes have one child on each side of the midline, where different rules apply. The result is chaos.

The structure of this tree begins to illuminate some of the observations made about the statistical distribution of Zeno game outcomes. For example, a simple counting argument explains the existence of gaps in the distribution. Every node of a binary tree has two descending links, which is just enough to reach all the dyadic rationals, whose population doubles at each level of the tree. But when links from two or more nodes converge on the same child node, then other nodes must be missing from the tree. (If I have two children and my spouse has two children, that doesn't necessarily mean we have four children.)

Counting can also explain why some game endings are more common than others, such as why $3/16$ turns up twice as often as $1/16$. Think of playing a Zeno game as tracing a pathway through the tree, starting at the root (level 0) and continuing down to some final node. At each interior node, the path turns left or right with equal probability. In the simple binary tree, this procedure yields the same probability for all the nodes at a given level. In particular, at the third level each of the eight nodes is reached with probability $1/8$. In the Zeno tree, however, two different paths both lead to the third-level node at $3/16$. Since there are two ways of getting to this node, the probability of landing there is doubled.

The path-counting analysis can be understood more clearly if we suppress some of the clutter and disorder in the Zeno tree. Suppose a random walk always takes steps of size $x/2$, adopting the rule that governs the left half of the Zeno tree but applying it everywhere. The corresponding tree has a uniform lattice structure in which it's easy to count paths (see figure 7.6). Every node has two parents, and the number of ways of reaching the node is simply the sum of the number of ways of reaching the parents. The path counts attached to the nodes of this tree may look familiar. They correspond to the entries in Pascal's triangle.

There's something else notable about this tidied-up version of the Zeno tree. All the numbers that appear in the lattice take a very specific form: $3^m/2^n$. Thus the numerators are drawn from the series 1, 3, 9, 27, 81,..., and the denominators follow the now-familiar progression 2, 4, 8, 16, 32,.... These are the very numbers that appear with higher-than-average frequency in the Zeno game, and it's clear why. The highest path counts run down the middle of the lattice, supporting the observation that numbers such as $3/32$, $9/64$, and $27/256$ are extraordinarily popular.

Of course, the real Zeno tree is not so tidy; the lattice is torn apart between any two nodes on opposite sides of the midline. As a result of these trans-midline events, factors other than 3 can enter the numerator; they form shadow lattices of their own, visible in some parts of the tree.

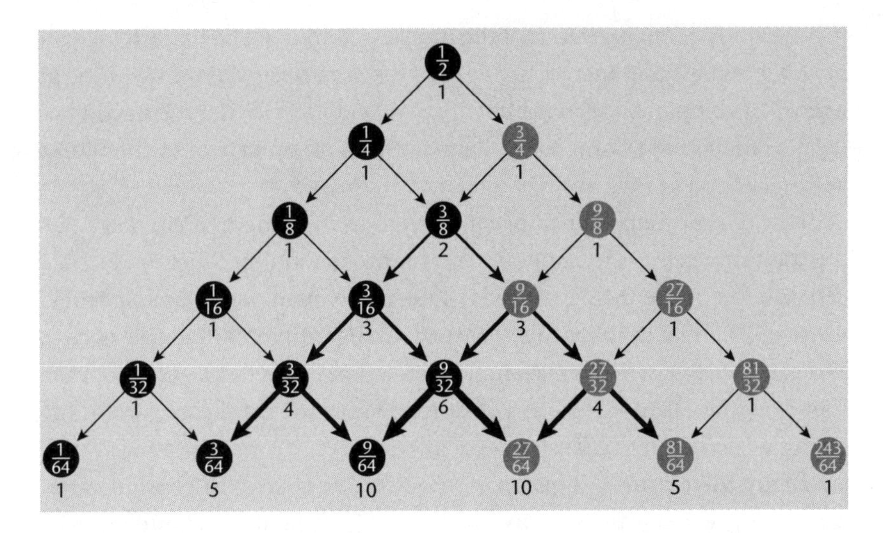

Figure 7.6 A simplified and regularized version of the Zeno game tree shows how junction points affect the number of paths through the tree and hence the probability of reaching a node. The tree corresponds to a random walk in which the step length is always one-half of x. The number of paths reaching each node (indicated by arrow thicknesses and by numbers in black) creates the pattern known as Pascal's triangle, with the likeliest nodes in the middle. The path counts in the real Zeno tree are different because the gray nodes at the right are not present in that tree.

Thus the probabilities calculated from the simplified lattice are at best approximations.

Midline Crises

That's all the answers I can offer on the Zeno wagering game, but I certainly haven't run out of questions. Here are three more.

First, how dense is the set of numbers included in the Zeno tree, measured as a proportion of all the numbers that *might* be present? At level n there are 2^n dyadic rationals in all; what fraction of them are Zeno tree numbers, and how does that fraction evolve as n increases? At the first three levels of the tree the fraction is 1; all the dyadic rationals are included. Then there is a steep linear descent as the fraction goes from $3/4$ to $5/8$ to $1/2$ to $3/8$. This linear series obviously cannot continue, or the tree would disappear entirely in three more steps. And indeed the slope decreases. The next four elements of the series are $9/32$, $13/64$, $19/128$, and $7/64$. At this point each level of the tree includes only about one tenth of the numbers that might be there. It seems a reasonable guess that the density will approach zero as n tends to infinity. In a series of computational experiments Carl Witty found that the fraction of dyadic rationals appearing in the Zeno tree falls off by a factor of about 0.72 with each level of the tree.

Second, how much structure can we find in the Zeno tree? To make this question more concrete, suppose you want to determine which node, on a certain level of the tree, gives the closest approximation to some specified value on the real number line. With the standard binary tree this is easy. You can specify a series of left and right turns that describe a path from the root to the closest node. It's not clear to me how to find such a path in the Zeno tree, except by exhaustive search.

Finally there's this big question: If you allow a game to continue arbitrarily long, will one player gain an advantage and hold onto it indefinitely, or will the lead change hands repeatedly? In terms of the random walk, will the walker get stuck on one side, or will the walk cross back and forth over the midline? It's always *possible* for the walker to get back to midline; all it takes is a sufficiently long sequence of steps in the right direction. But it doesn't necessarily follow that such an event will have a probability greater than zero when the number of steps grows without limit.

The experimental evidence on this point leaves little room for doubt. In random walks of 10,000 steps, roughly half exhibit no midline crossings at all. The walker sets out either left or right initially and never gets back to the other side. For those walks that do cross the midline, the last observed crossing is usually within the first ten steps; the latest I have seen is step 101. After 10,000 steps the probability of another midline crossing is clearly very small. But does the probability go to zero as the number of steps goes to infinity?

I was not able to answer this question on my own, but Scott Aaronson, now of the University of Texas at Austin, quickly showed that the Zeno walk has the same limiting behavior as a simpler process whose long-term fate is easier to determine. In the simplified model a walker goes either one step toward the midline or C steps away from it with equal probability. The constant C is an aribtrary number greater than 1. Because of the bias favoring movement away from the midline, Aaronson's walker can escape to unbounded distances. The probability of returning to the origin falls exponentially, approaching a limiting value of zero as the number of steps increases. Although this model differs from the Zeno walk in crucial details—for example, the walker ranges over the entire number line, not a finite segment—the limiting probability is the same.

In Search of Zeno's Urn

The literature of probability theory offers a huge trove of other models that might illuminate the Zeno process. Knowledgeable friends have pointed me to two areas in particular: urn models and reinforced random walks. I have not found any discussion of a model that I can recognize as being isomorphic to the Zeno game, but I've found some fascinating reading along the way.

Urn models are the eighteenth-century ancestor of the Powerball lottery. You mix up a bunch of balls in a container and draw them out one by one. Of particular relevance is a class of models studied in the 1920s by George Pólya, later by Bernard Friedman and David A. Freedman and others, and more recently by Robin Pemantle. In one version, an urn initially holds one white ball and one black ball. Then each time a ball is drawn at random from the urn, that ball is put back in along with an extra ball of the same color. When the process is repeated many times, will the ratio of black to white balls settle down to some fixed value, or will it keep fluctuating? The answer is that in any single long run, the ratio will tend to a stable value, but the value itself is a random variable. If you try the experiment again, it will come out differently.

The reinforced random walk was invented by Persi Diaconis and has also been studied by Pemantle. The idea is to watch a random walker stepping from node to node through a network. Visiting a node increases the probability that the same node will be chosen the next time the walker is nearby. The question asked is whether the walker can range freely throughout the graph forever or will get trapped in some local neighborhood. The answer seems to depend on the details. On a one-dimensional lattice (like the line of integers), the walker gets stuck in a five-node region. But a variation that associates probabilities with the links between nodes rather than the nodes themselves allows the walker to escape confinement and visit every node infinitely often. On two-dimensional infinite lattices the fate of the walker is unknown.

The Zeno game seems to belong in the same general family as these models. Although the Zeno mechanism includes no explicit reinforcement of probabilities, it has an indirect form of positive feedback, because movement away from the center reduces the step size and thereby makes it harder to move back. Perhaps the Zeno process can be reformulated in a way that will make the correspondence with known work clearer. But the analysis of such models is a subtle art. Every time I get close to an answer, it seems the problem has moved on a little further down the road.

8

The Higher Arithmetic

In 2008 the National Debt Clock ran out of digits. The billboard-size electronic counter, mounted on a wall near Times Square in New York City, overflowed when the public debt reached \$10 trillion, or 10^{13} dollars. The crisis was resolved by squeezing another digit into the space occupied by the dollar sign. As I write this in late 2016 the tally is approaching \$20 trillion.

The incident of the Debt Clock brings to mind a comment made by Richard Feynman in the 1980s—back when mere billions still had the power to impress:

> There are 10^{11} stars in the galaxy. That used to be a *huge* number. But it's only a hundred billion. It's less than the national deficit! We used to call them astronomical numbers. Now we should call them economical numbers.

The important point here is not that high finance is catching up with the sciences; it's that the numbers we encounter everywhere in daily life are growing steadily larger. Computer technology is another area of rapid numeric inflation. Data storage capacity has gone from kilobytes to megabytes to gigabytes to terabytes (10^{12} bytes). In the world of supercomputers, the current state of the art is called petascale computing (10^{15} operations per second), and there is much talk of a coming transition to exascale (10^{18}). After that, we can await the arrival of zettascale (10^{21}) and yottascale (10^{24}) machines—and then we run out of prefixes!

Even these numbers are puny compared with the prodigious creations of pure mathematics. In the eighteenth century the largest known prime number had 10 digits; the present record-holder runs to 22 million digits. The value of π has been calculated to a trillion digits—a feat at once magnificent and mind-numbing. Elsewhere in mathematics there are numbers so big that even trying to describe their size requires numbers that are too big to describe. Of course, none of these numbers are likely to turn up in everyday chores such as balancing a checkbook. On the other hand, logging into a bank's website involves doing arithmetic with numbers in the vicinity of 2^{128}, or 10^{38}. (The calculations take place behind the scenes, in the cryptographic protocols meant to ensure privacy and security.)

Which brings me to the theme of this essay. Those streams of digits that make us so dizzy also present challenges for the design of computer hardware and software. Like the National Debt Clock, computers often set rigid limits on the size of numbers. When routine calculations begin to bump up against those limits, it's time for a rethinking of numeric formats and algorithms. For more than 50 years most numerical calculations have been done with a technology called floating-point arithmetic. It's not going away, but there may be room for alternative schemes for computing with astronomical and economical and mathematical numbers.

Numerical Eden

In their native habitat—which is *not* the digital computer—numbers are boundless and free-ranging. Along the real number line are infinitely many integers, or whole numbers. Between any two integers are infinitely many rational numbers, such as $3/2$ and $5/4$. Between any two rationals are infinitely many irrationals—numbers like $\sqrt{2}$ or π.

The reals are a Garden of Eden for doing arithmetic. Just follow a few simple rules—such as not dividing by zero—and these numbers will never lead you astray. They form a safe, closed universe. If you start with any set of real numbers, you can add and subtract and multiply all day long—and divide, too, except by zero—and at the end you'll still have real numbers. There's no risk of slipping through the cracks or going out of bounds.

Unfortunately, digital computers exist only outside the gates of Eden. Out here, arithmetic is a treacherous process. Even simple counting can get you in trouble. With computational numbers, adding 1 over and over eventually brings you to a largest number—something unknown in mathematics. If you try to press on beyond this limit, there's no telling what will happen. The next number after the largest number might be the smallest number; or it might be something labeled ∞; or the machine might sound an alarm or die in a puff of smoke.

This is a lawless territory. On the real number line, you can always rely on principles like the associative law: $(a+b)+c=a+(b+c)$. In some versions of computer arithmetic, that law breaks down. (Try it with $a = 10^{30}$, $b = -10^{30}$, $c = 1$.) And when calculations include irrational numbers—well, irrationals just don't exist in the digital world. They have to be approximated by rationals—the very thing they are defined not to be. As a result, mathematical identities such as $(\sqrt{2})^2 = 2$ are not to be trusted.

Bignums

The kind of computer arithmetic that comes closest to the mathematical ideal is calculation with integers and rationals of arbitrary size, limited only by the machine's memory capacity. In this bignum arithmetic, an integer is stored as a long sequence of bits, filling up as much space as needed. A rational number is a pair of such integers, interpreted as a numerator and a denominator.

A few primitive computers from the vacuum-tube era had built-in hardware for doing arithmetic on integers of arbitrary size, but our sophisticated modern machines have lost that capability, and so the process has to be orchestrated by software. Adding two integers proceeds piece by piece, starting with the least significant bits and working right to left, much as a paper-and-pencil algorithm sums pairs of digits one at a time, propagating any carries to the next column. The usual practice is to break up the sequence of bits into blocks the size of a machine register—typically 32 or 64 bits. Algorithms for multiplication and division follow similar principles; operations on rationals require the further step of reducing a fraction to lowest terms.

Looking beyond integers and rationals, there have even been efforts to include irrational numbers in exact computations. Of course, there's no hope of expressing the complete value of π or $\sqrt{2}$ in a finite machine, but a program can calculate the values incrementally, supplying digits as they are needed—a strategy known as lazy computing. For example, the assertion $\pi < 3.1414$ could be tested—and shown to be false—by generating the first five decimal digits of π. Another approach is to treat irrational numbers as unevaluated units, which are carried through the computation from start to finish as symbols; thus the circumference of a circle of unit radius would be given simply as 2π.

The great virtue of bignum arithmetic is exactness. If the machine ever gives an answer, it will be the right answer (barring bugs and hardware failures). But there's a price to pay: you may get no answer at all. The program could run out of memory, or it could take so long that it exhausts human patience or the human life span.

For some computations, exactness is crucial, and bignum arithmetic is the only suitable choice. If you want to search for million-digit primes, you have to look at every last digit. Similarly, the security module in a web browser must work with the exact value of a cryptographic key.

For many other kinds of computations, however, exactness is neither needed nor helpful. Using exact rational arithmetic to calculate the interest on a mortgage loan yields an unwieldy fraction accurate to hundreds of decimal places, but knowing the answer to the nearest penny would suffice. In many cases the inputs to a computation come from physical measurements accurate to no more than a few significant digits; lavishing exact calculations on these measurements cannot make them any more accurate.

What's the Point?

Most computer arithmetic is done not with bignums or exact rationals but with numbers confined to a fixed allotment of space, such as 32 or 64 bits. The hardware operates on all the bits at once, so arithmetic can be very fast. But an implacable law governs all such fixed-size formats. If a number is represented by 32 bits, then it can take on at most 2^{32} possible values. You may be able to choose *which* 2^{32} values are included, but there's no way to increase the size of the set.

For 32-bit numbers, one obvious mapping assigns the 2^{32} bit patterns to the integers from 0 through 4,294,967,295 (which is $2^{32}-1$). The same range of integers could be shifted along the number line, or the values could be scaled to cover a smaller numerical range in finer increments (perhaps 0.00 up to 42,949,672.95) or spread out over a wider range more sparsely. Arithmetic done in this style is known as fixed-point, since the position of the decimal point is the same in all numbers of a given class.

Fixed-point arithmetic was once the mainstay of numerical computing, and it still has a place in certain applications, such as high-speed signal processing. But the dominant format now is floating-point, where the decimal point (or binary point) can be moved around to represent a wide range of magnitudes. The floating-point format is based on the same idea as scientific notation. Just as we can write a large number succinctly as 6.02×10^{23}, floating-point arithmetic stores a number in two parts: the significand (6.02 in this example) and the exponent (23).

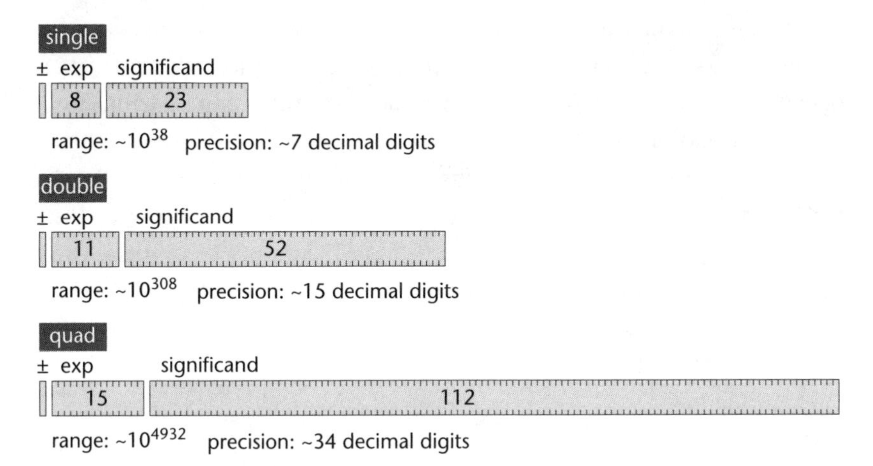

Figure 8.1 Floating-point numbers are reminiscent of scientific notation. A quantity is represented by a significand, an exponent, and a sign bit. Three floating-point formats included in the current standard have total widths of 32, 64, and 128 bits. The "single" format, for example, allocates 1 bit for the sign, 8 bits for the exponent, and 23 bits for the significand.

Designing a floating-point format entails a compromise between range and precision (see figure 8.1). Every bit allocated to the significand doubles its precision, but the bit has to be taken from the exponent, and it therefore reduces the range by half. For 32-bit numbers the prevailing standard dictates a 23-bit significand and an 8-bit exponent, with 1 bit reserved for the sign of the significand. Representable numbers range from 2^{-126} up to 2^{127}; in decimal notation the largest number is about 3×10^{38}. Standard 64-bit numbers allocate 52 bits to the significand and 11 to the exponent, allowing a range up to about 10^{308}.

The idea of floating-point arithmetic goes back to the beginning of the computer age, starting with Konrad Zuse's pioneering electromechanical machines circa 1940. Floating-point hardware was an optional extra for several early mainframes, but every implementation was a little different. An impetus to wider adoption was the drafting of a standard, approved by the Institute of Electrical and Electronics Engineers (IEEE) in 1985. This effort was led by William Kahan of the University of California, Berkeley, who remains a strong advocate of the technology.

Early critics of the floating-point approach worried about efficiency and complexity. In fixed-point arithmetic, many operations can be reduced to a single machine instruction, but floating-point calculations are more involved. First you have to extract the significands and exponents, then operate on these pieces separately, then do some rounding and adjusting, and finally reassemble the parts.

The answer to these concerns was to implement floating-point algorithms in hardware. Even before the IEEE standard was approved, Intel designed a floating-point coprocessor for early personal computers. Later generations incorporated a floating-point unit on the main processor chip. From the programmer's point of view, floating-point arithmetic became part of the infrastructure.

Safety in Numbers

It's tempting to pretend that floating-point arithmetic is simply real-number arithmetic in silicon. This attitude is encouraged by programming languages that use the label *real* for floating-point variables. But floating-point numbers are *not* real numbers; at best they provide a finite model of the infinite real number line.

Unlike the real numbers, the floating-point universe is not a closed system. When you multiply two floating-point numbers, there's a good chance that the product—the *real* product, as calculated in real arithmetic—will not be a floating-point number. This leads to three kinds of problems.

The first problem is rounding error. A number that falls between two floating-point values has to be rounded by shifting it to one or the other of the nearest representable numbers. The resulting loss of accuracy is usually small and inconsequential, but circumstances can conspire to produce numerical disasters. A notably risky operation is subtracting one large quantity from another, which can wipe out all the significant digits in the small difference. Textbooks on numerical analysis are heavy with advice on how to guard against such events; mostly it comes down to, "Don't do that."

The second problem is overflow, when a number goes off the scale. The IEEE standard allows two responses to this situation. The computer can halt the computation and report an error, or it can substitute a special marker, "∞," for the oversize number. The latter option is designed to mimic the properties of mathematical infinities; for example, $\infty + 1 = \infty$. Because of this behavior, floating-point infinity is a black hole. Once you get into it, there is no way out, and all information about where you came from is annihilated.

The third hazard is underflow, where a number too small to represent collapses to zero. In real arithmetic, a sequence like ½, ¼, ⅛,... can go on indefinitely, but in a finite floating-point system there must be a smallest nonzero number. On the surface, underflow looks much less serious than overflow. After all, if a number is so small that the computer can't distinguish it from zero, what's the harm in making it exactly zero? But this reasoning is misleading. In the exponential space of floating-point numbers, the distance from, say, 2^{-127} to zero is exactly the same as the distance from 2^{127} to infinity. As a practical matter, underflow is a frequent cause of failure in numerical computations.

Problems of rounding, overflow, and underflow cannot be entirely avoided in any finite number system. They can be ameliorated, however, by adopting a format with higher precision and a wider range—by throwing more bits at the problem. This is one approach taken in a revision of

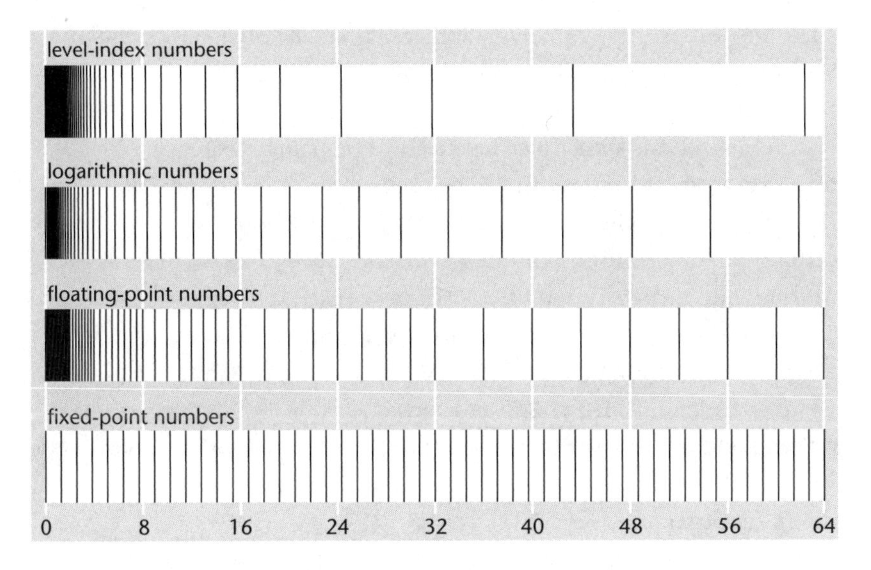

Figure 8.2 "Spectra" of computer number systems show how numbers are distributed along the real number line. Fixed-point numbers are placed at uniform intervals; for floating-point numbers the density falls by half with each higher power of 2; logarithmic numbers have a smoothly declining density; so do level-index numbers, but the gradient in density is even more extreme. The spectra are based on toy versions of the number systems, with just a few bits of precision.

the IEEE standard, approved in 2008. It includes a 128-bit floating-point format supporting numbers as large as 2^{16383}, or about 10^{4932}, and precision has increased to about 34 decimal digits. (But no widely available hardware yet supports this quad-precision format directly.)

Tapering Off, or Rolling Off a Log

By now, IEEE floating-point methods are so firmly established that they often seem like the *only* way to do arithmetic with a computer. But many alternatives have been discussed over the years (see figure 8.2). Here I describe two of them briefly and take a somewhat closer look at a third idea.

The first family of proposals might be viewed more as an enhancement of floating-point than as a replacement. The idea is to make the trade-off between precision and range an adjustable parameter. If a calculation does not require very large or very small numbers, then it can give more bits to the significand. Other programs might want to sacrifice precision in order to gain wider scope for the exponent. To make such flexibility possible, it's necessary to set aside a few bits to keep track of how the other bits are allocated. (Of course, those bookkeeping bits are thereby made unavailable for either the exponent or the significand.)

A scheme of this kind, called tapered floating-point, was proposed in 1971 by Robert Morris, who was then at Bell Laboratories. A decade later, more elaborate plans were published by Shouichi Matsui and Masao Iri of the University of Tokyo and by Hozumi Hamada of Hitachi, Ltd. In 2006 Alan Feldstein of Arizona State University and Peter R. Turner of Clarkson University described a tapered scheme that works exactly like a conventional floating-point system except when overflow or underflow threaten. And John L. Gustafson proposed a numeric scheme called unums in which one can adjust the sizes of the significand and the exponent and even the size of the fields that specify those sizes. (Gustafson presents his ideas in a book whose title makes quite a promise: *The End of Error*.)

The second alternative to the floating-point system would replace numbers by their logarithms. For example, in a decimal version of the plan the number 751 would be stored as 2.87564, since $10^{2.87564} = 751$. This plan is not as radical a departure as it might see because floating-point is already a semi-logarithmic notation; the exponent of a floating-point number is the integer part of a logarithm. Thus the two formats record essentially the same information.

If the systems are so similar, what's gained by the logarithmic alternative? The motive is the same as that for developing logarithms in the first place:. They facilitate multiplication and division, reducing those operations to addition and subtraction. For positive numbers a and b, $\log(ab) = \log(a) + \log(b)$. In general, multiplying takes more work than adding, so this substitution is a net gain. But there's another side to the coin. Although logarithms make multiplying easy, they make adding hard. Computing $a+b$ when you have only $\log(a)$ and $\log(b)$ is not straightforward. For this reason logarithmic arithmetic is attractive mainly in specialized areas, such as image processing, where multiplications tend to outnumber additions.

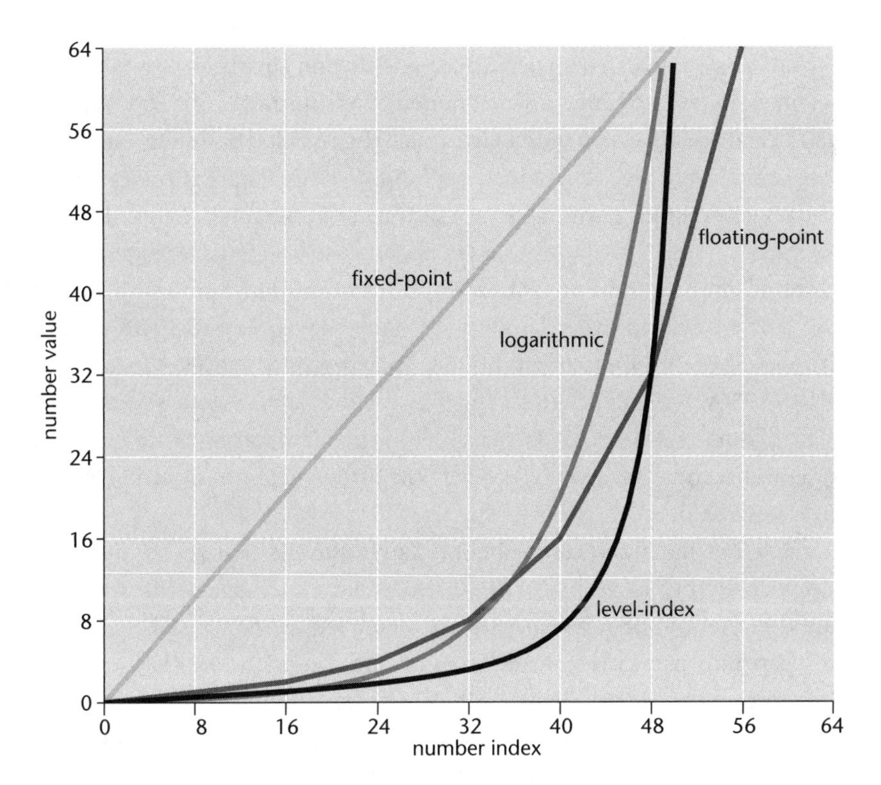

Figure 8.3 Another visualization of number system spectra shows how the magnitude of a number increases as a function of the number's position in the counting sequence. For the uniformly spaced fixed-point numbers, the function is a straight line, but the other systems produce concave upward curves (or, in the case of floating-point, a jointed sequence of straight segments). The level-index system has the highest density of small numbers, then the steepest rate of growth for larger ones.

On the Level

The third scheme I want to mention here addresses the problem of overflow. If you are trying to maximize the range of a number system, an idea that pops up naturally is to replace mere exponents with towers of exponents. If 2^n can't produce a number large enough for your needs, then try

$$2^{2^n} \quad \text{or} \quad 2^{2^{2^n}} \quad \text{or} \quad 2^{2^{2^{2^n}}}.$$

(Whatever the mathematical merits of such expressions, they are a typographical nightmare, so from here on I adopt a more convenient notation, invented by Donald E. Knuth of Stanford University: $2\uparrow2\uparrow2\uparrow2\uparrow n$ is equivalent to the last of those three towers. It is to be evaluated from right to left, just as the tower is evaluated from top to bottom.)

Number systems based on iterated exponentiation have been proposed several times; for example, they are mentioned by Matsui and Iri and by Hamada. But one particular version of the idea, called the level-index system, has been worked out with such care and thoughtful analysis that it deserves closer attention. Level-index arithmetic is a lost gem of computer science. It may never make it into the CPU of your laptop, but it shouldn't be forgotten.

The scheme was devised by Charles W. Clenshaw and Frank W. J. Olver, who first worked together (along with Alan Turing) in the 1940s at the National Physical Laboratory in Britain. They proposed the level-index idea in the 1980s, writing a series of papers on the subject with several colleagues, notably Daniel W. Lozier, now of the National Institute of Standards and Technology (NIST), and Peter R. Turner. (Clenshaw died in 2004, Olver in 2013.)

Iterated exponentials can be built on any numeric base; most proposals have focused on base 2 or base 10. Clenshaw and Olver argued that the best base is e, the irrational number usually described as the base of the natural logarithms or as the limiting value of the compound interest formula $(1+\frac{1}{n})^n$; numerically e is about 2.71828. Building numbers on an irrational base is an idea that takes some getting used to. For one thing, it means that almost all numbers that have an exact representation are irrational; the only exceptions are 0 and 1. But there's no theoretical difficulty in constructing such numbers, and there's a good reason for choosing base e.

In the level-index system a number is represented by an expression of the form $e\uparrow e\uparrow \ldots \uparrow e\uparrow m$, where the m at the end of the chain is a fractional quantity analogous to the mantissa of a logarithm. The number of up arrows—in other words, the height of the exponential tower—depends on the magnitude of the number being represented.

To convert a positive number to level-index form, we first take the logarithm of the number, then the logarithm of the logarithm, and so on, continuing until the result lies in the interval between 0 and 1. Counting the successive logarithm operations gives us the *level* part of the representation; the remaining fraction becomes the *index*, the value of m in the preceding expression. The process is defined by the function $f(x)$:

```
if 0 ≤ x < 1
   then f(x) = x
   else f(x) = 1 + f(log(x))
```

Here's how the procedure applies to a national debt amount:

$$
\begin{aligned}
\log(19{,}862{,}390{,}036{,}870) &= 30.619848,\\
\log(30.619848) &= 3.4216485,\\
\log(3.4216485) &= 1.2301224,\\
\log(1.2301224) &= 0.2071137.
\end{aligned}
$$

We've taken logarithms four times, so the level is 4, and the fractional amount remaining becomes the index. Thus the level-index form of the national debt is 4.2071137 (which seems a lot less worrisome than \$19,862,390,036,870).

The level-index system accommodates *very* large numbers. Level 0 runs from 0 to 1, then level 1 includes all numbers up to e. Level 2 extends as far as $e{\uparrow}e$, or about 15.2. Beyond this point, the growth rate gets steep. Level 3 goes up to $e{\uparrow}e{\uparrow}e$, which is about 3,814,273. Continuing the ascent through level 4, we soon pass the largest 64-bit floating-point number, which has a level-index value of about 4.63. The upper boundary of level 4 is a number with 1.6 million decimal digits. Climbing higher still puts us in the realm of numbers where even a description of the size is hopelessly impractical. Just seven levels are enough to represent all distinguishable level-index numbers. Thus only three bits need to be devoted to the level; the rest can be used for the index.

What about the other end of the number scale—the very small numbers? The level-index system is adequate for many purposes in this region, but a variation called symmetric level index provides additional precision close to zero. In this scheme a number x between 0 and 1 is denoted by the level-index representation of $1/x$.

Apart from its wide range, the level-index system has some other distinctive properties. One is smoothness. For floating-point numbers, a graph of the magnitudes of successive numbers is a jointed sequence of straight lines, with an abrupt change of slope at each power of 2 (see figure 8.3). The corresponding graph for the level-index system is a smooth curve. For iterated exponentials this is true only in base *e*, which is the reason for choosing that base.

Olver also pointed out that level-index arithmetic is a closed system, like arithmetic with real numbers. How can that be? Since level-index numbers are finite, there must be a largest member of the set, and so repeated additions or multiplications should eventually exceed that bound. Although this reasoning is unassailable, it turns out that the system does not in fact overflow. Here's what happens instead. Start with a number x, then add or multiply to generate a new larger x, which is rounded to the nearest level-index number. As x grows very large, the available level-index values become sparse. At some point, the spacing between successive level-index values is greater than the change in x caused by addition or multiplication. Thereafter, successive iterations of x round to the same level-index value.

This is not a perfect model of unbounded arithmetic. In particular, the process is not reversible. A long series of $x+1$ operations followed by an equal number of $x-1$s will not bring you back to where you started, as it would on the real number line. Still, the boundary at the end of the number line seems about as natural as it can be in a finite system.

Shaping a Number System

Is there any genuine need for an arithmetic that can reach beyond the limits of IEEE floating-point? I have to admit that I seldom write a program whose output is a number greater than 10^{38}. But that's not the end of the story.

A program with inputs and outputs of only modest size may nonetheless generate awkwardly large intermediate values. Suppose you want to know the probability of observing exactly 1,000 heads in 2,000 tosses of a fair coin. The standard formula calls for evaluating the factorial of 2,000, which is $1 \times 2 \times 3 \times \ldots \times 2{,}000$ and is sure to overflow. You also need to calculate $(\frac{1}{2})^{2000}$, which could underflow. Although the computation *can* be successfully completed with floating-point numbers—the answer is about 0.018—it requires careful attention to cancellations and reorderings of the operations. A number system with a wider range would allow a simpler and more robust approach.

In 1993 Lozier described a more substantial example of a program sensitive to numerical range. A simulation in fluid dynamics failed because of severe floating-point underflow; redoing the computation with the symmetric version of level-index arithmetic produced correct output.

Persuading the world to adopt a new kind of arithmetic is a quixotic undertaking, like trying to reform the calendar or replace the QWERTY keyboard. But even setting aside all the obstacles of history and habit, I'm not sure how best to evaluate the alternatives in this case. The main conceptual question is this: Since we don't have enough numbers to cover the entire number line, what is the best distribution of the numbers we *do* have? Fixed-point systems sprinkle them uniformly. Floating-point numbers are densely packed near the origin and grow farther apart out in the numerical hinterland. In the level-index system, the core density is even greater, and it drops off even more steeply, allowing the numbers to reach the remotest outposts.

Which of these distributions should we prefer? Perhaps the answer will depend on what numbers we need to represent—and thus on how quickly the national debt continues to grow.

9

First Links in the Markov Chain

It was an unlikely partnership of poetry and probability theory. Delving into the text of Alexander Pushkin's novel-in-verse *Eugene Onegin,* the Russian mathematician A. A. Markov spent hours sifting through patterns of vowels and consonants. On January 23, 1913, he summarized his findings in an address to the Imperial Academy of Sciences in St. Petersburg. His analysis did not alter the understanding or appreciation of Pushkin's poem, but the technique he developed—now known as a Markov chain—extended the theory of probability in a new direction. Markov's methodology went beyond the familiar coin-flipping and dice-rolling models that had long been the mainstay of probability studies; in those situations each event is independent of all the others. Markov introduced chains of linked events, where what happens next depends on what just happened.

Markov chains are everywhere in the sciences today. Methods not too different from those Markov used in his study of Pushkin help identify genes in DNA, and they power algorithms for voice recognition and web search. In physics the Markov chain simulates the collective behavior of systems made up of many interacting particles, such as the electrons in a solid. In statistics the chains provide methods of drawing a representative sample from a large set of possibilities. They've been used for optimizing the batting order on baseball teams. And within mathematics Markov chains themselves have become a lively area of inquiry in recent decades, with efforts to understand why some of them work so efficiently—and some don't.

As Markov chains have become commonplace tools, the story of their origin has faded from memory. That story is worth retelling. It features an unusual conjunction of mathematics and literature as well as a bit of politics and even theology. For added drama there's a bitter feud between two forceful personalities. And the story unfolds amid the tumultuous events that transformed Russian society in the early years of the twentieth century.

Before delving into the early history of Markov chains, however, it's helpful to have a clearer idea of what the chains are and how they work.

A Markovian Weather Forecast

Probability theory has its roots in games of chance, where every roll of the dice or spin of the roulette wheel is a separate experiment, independent of all others. It's an article of faith that one flip of a coin has no effect on the next. If the coin is fair, the probability of heads is always $\frac{1}{2}$.

This principle of independence makes it easy to calculate compound probabilities. If you toss a fair coin twice, the chance of seeing heads both times is simply $\frac{1}{2} \times \frac{1}{2}$, or $\frac{1}{4}$. More generally, if two independent events have probabilities p and q, the joint probability of both events is the product pq.

However, not all aspects of life adhere to this convenient principle. Suppose the probability of a rainy day is $\frac{1}{3}$; it does *not* follow that the probability of rain two days in a row is $\frac{1}{3} \times \frac{1}{3} = \frac{1}{9}$. Storms often last several days, so rain today may signal an elevated chance of rain tomorrow.

For another example where independence fails, consider the game of Monopoly. Rolling the dice determines how many steps your token advances around the board, but where you land at the end of a move obviously depends on where you begin. From different starting points, the same number of steps could take you to the Boardwalk or put you in jail. The probabilities of future events depend on the current state of the system. The events are linked, one to the next; they form a Markov chain.

To be considered a proper Markov chain, a system must have a set of distinct states, with identifiable transitions between them. A simplified model of weather forecasting might have just three states: *sunny, cloudy,* and *rainy* (see figure 9.1). There are nine possible transitions (including identity transitions that leave the state unchanged). For Monopoly, the minimal model would require at least 40 states, corresponding to the 40 squares around the perimeter of the board. For each state there are transitions to all other states that can be reached in a roll of the dice—generally those from 2 to 12 squares away. A realistic Monopoly model incorporating all the game's quirky rules would be much larger.

Recent years have seen the construction of truly enormous Markov chains. For example, the PageRank algorithm devised by Larry Page and

Sergey Brin, the founders of Google, is based on a Markov chain whose states are the pages of the World Wide Web—40 or 50 billion of them. The transitions are hypertext links between pages. The aim of the algorithm is to calculate for each web page the probability that a reader following links at random will arrive at that page.

Past, Present, and Future

A diagram made up of dots and arrows shows the structure of a Markov chain. Dots represent states; arrows indicate transitions. Each arrow has an associated number, which gives the probability of that transition. Because these numbers are probabilities, they must lie between 0 and 1, and all the probabilities issuing from a dot must add up to exactly 1. In such a diagram you can trace a pathway that defines a sequence of states—perhaps *sunny, sunny, cloudy, rainy* in the weather example. To calculate the probability of this specific sequence, just multiply the probabilities associated with the corresponding transition arrows (see figure 9.2).

The chain can also answer questions such as, If it's cloudy today, what is the probability of rain two days from now? The answer is found by summing the contributions of all pathways that lead from the *cloudy* state to

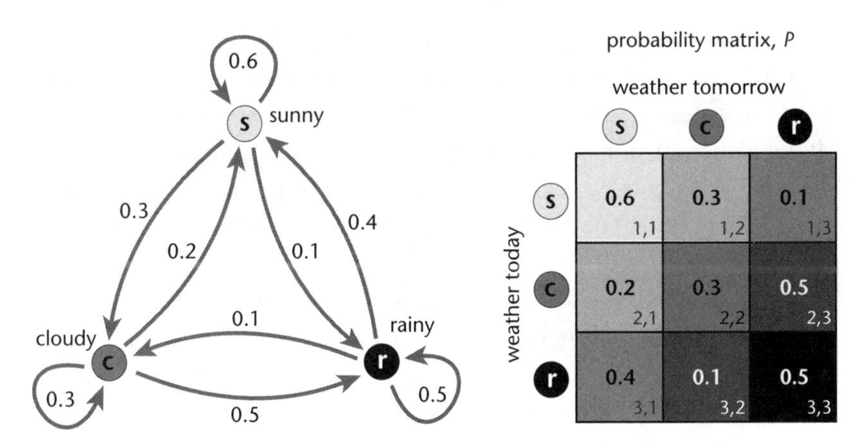

Figure 9.1 A Markov chain describes a set of states and transitions between them. In the diagram at left the states of a simple weather model are represented by dots labeled *s* for sunny, *c* for cloudy, and *r* for rainy; transitions between the states are indicated by arcs, or arrows, each of which has an associated probability. The probabilities can be arranged in a 3×3 matrix *(right)*, with each row giving one possible state of today's weather and each column a corresponding state tomorrow.

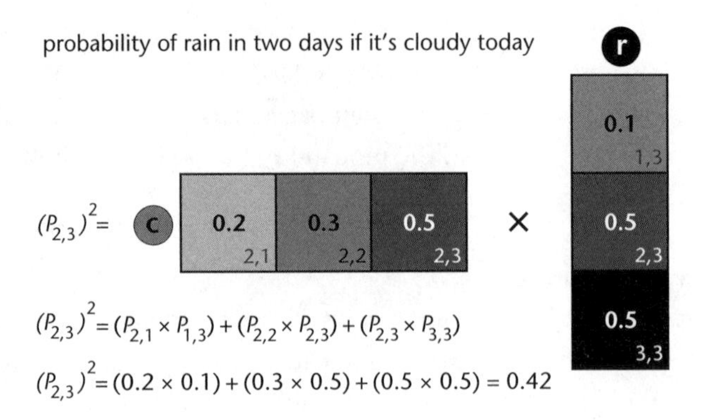

Figure 9.2 A weather forecast for the day after tomorrow in the Markov model is a matter of matrix multiplication—specifically, multiplying the matrix by itself. The details of the multiplication algorithm for one element of the matrix are shown here. If it's cloudy today, the probability of rain in two days is found by multiplying the elements of the *cloudy* row by those of the *rainy* column, then summing the three products.

the *rainy* state in exactly two steps. This sounds like a tedious exercise, but there's an easy way to organize the computation, based on the arithmetic of matrices.

The transition probabilities for a three-state Markov chain can be arranged in a 3×3 matrix—a square array of nine numbers (see figure 9.3). The process for calculating multistage transitions is equivalent to matrix multiplication. The matrix itself (call it P) predicts tomorrow's weather; the product $P \times P$, or P^2, gives weather probabilities for the day after tomorrow; P^3 defines the probabilities for three days hence, and so on. The entire future unfolds from this one matrix.

Given these hypothetical weather probabilities, the successive powers of the matrix rapidly converge to a stationary configuration in which all the rows are identical and all the columns consist of a single repeated value. This outcome has a straightforward interpretation. If you let the system evolve long enough, the probability of a given state no longer depends on the initial state. In the case of the weather, such a result is unsurprising. Knowing that it's raining today may offer a clue about tomorrow's weather, but it's not much help in predicting the state of the skies three weeks from now. For such an extended forecast you may as well consult the long-term averages (which are the values to which the Markov chain converges).

Markov's scheme for extending the laws of probability beyond the realm of independent variables has one crucial restriction: The probabilities must depend only on the present state of the system, not on its earlier history. The Markovian analysis of Monopoly, for example, considers a player's current position on the board but not how he or she got there. This limitation is serious. After all, life presents itself as a long sequence of contingent events—kingdoms are lost for want of a nail, hurricanes are spawned by butterflies in Amazonia—but these causal chains extending into the distant past are not Markov chains.

On the other hand, a finite span of history can often be captured by encoding it in the current state. For example, tomorrow's weather could be made dependent on both yesterday's and today's by creating a nine-state model in which each state is a two-day sequence. The price to be paid is an exponential increase in the number of states.

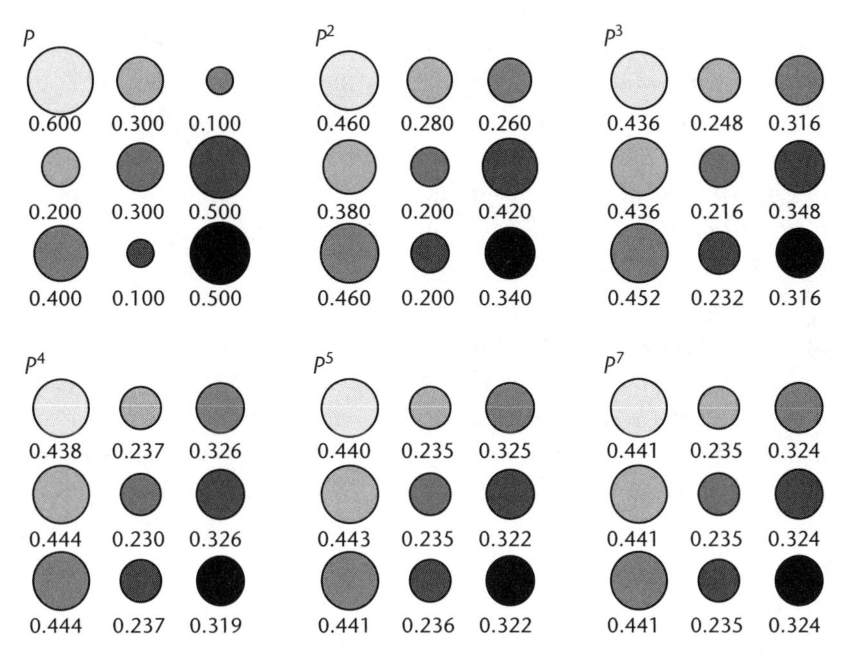

Figure 9.3 Raising the matrix to higher powers gives probabilities of longer sequences of transitions between states. After just a few such steps, the matrix begins to converge to a stationary configuration where all the rows are identical and all the columns consist of a single repeated value. In each row the probabilities continue to sum to 1.

Peterburgian Math

In trying to understand how Markov came to formulate these ideas, we run straight into one of those long chains of contingent events extending deep into the past. One place to start the story is with Peter the Great (1672–1725), the ambitious Romanov tsar who founded the Academy of Sciences in St. Petersburg and fostered the development of scientific culture in Russia. (Other aspects of his reign were less admirable, such as the torture and murder of dissidents, including his son Alexei.)

At roughly the same time, elsewhere in Europe, the theory of probability was emerging from gambling halls and insurance brokerages to become a coherent branch of mathematics. The foundational event was the publication in 1713 of Jacob Bernoulli's treatise *Ars Conjectandi* (*The Art of Conjecturing*).

Back in St. Petersburg, the mathematics program prospered, although initially most of the accomplishments were by imported savants. Visitors to the Academy included two younger members of the Bernoulli family, Nicholas and Daniel. And the superstar was Leonhard Euler, the preeminent mathematician of the era, who spent more than 30 years in St. Petersburg.

By the nineteenth century indigenous Russian mathematicians were beginning to make their mark. Nikolai Lobachevsky (1792–1856) was one of the inventors of non-Euclidean geometry. A few decades later, Pafnuty Chebyshev (1821–1894) made contributions in number theory, in methods of approximation (now called Chebyshev polynomials), and in probability theory. Chebyshev's students formed the nucleus of the next generation of Russian mathematicians; Markov was prominent among them.

Andrei Andreevich Markov was born in 1856. His father was a government employee in the forestry service and later the manager of an aristocrat's estate. As a schoolboy Markov showed enthusiasm for mathematics. He went on to study at St. Petersburg University (with Chebyshev and others) and remained there for his entire career, progressing through the ranks and becoming a full professor in 1893. He was also elected to the Academy.

In 1906, when Markov began developing his ideas about chains of linked probabilities, he was 50 years old and had already retired, although he still taught occasional courses. His retirement was active in another way as well. In 1883 Markov had married Maria Valvatieva, the daughter of the owner of the estate his father had once managed. In 1903 they had their first and only child, a son who was also named Andrei Andreevich. The son became a distinguished mathematician of the Soviet era, head of the department of mathematical logic at Moscow State University. (To the consternation of librarians, both father and son signed their works "A. A. Markov.")

An intellectual thread extends all the way from Jacob Bernoulli through Chebyshev to the elder Markov. In *Ars Conjectandi* Bernoulli stated the law of large numbers, which says that if you keep flipping an unbiased coin, the proportion of heads will approach ½ as the number of flips goes to infinity. This notion seems intuitively obvious, but it gets slippery when you try to state it precisely and supply a rigorous proof. Bernoulli proved one version; Chebyshev published a broader proof; Markov offered further refinements.

Markov's later studies of chains of dependent events can be seen as a natural continuation and generalization of this long line of work. But that's not the whole story.

Mathematical Theology

By most accounts, Markov was a nettlesome character, abrasive even with friends, fiercely combative with rivals, often embroiled in public protests and quarrels. We get a glimpse of his personality from his correspondence with the statistician Alexander Chuprov, which has been published in English translation. His letters to Chuprov are studded with dismissive remarks denigrating others' work, including Chuprov's.

Markov's pugnacity extended beyond mathematics to politics and public life. When the Russian church excommunicated Leo Tolstoy, Markov asked that he be expelled, also. (The request was granted.) In 1902 the leftist writer Maxim Gorky was elected to the Academy, but the election was vetoed by Tsar Nicholas II. In protest, Markov announced that he would refuse all future honors from the tsar. (Unlike Anton Chekhov, however, Markov did not resign his own membership in the Academy.) In 1913, when the tsar called for celebrations of 300 years of Romanov rule, Markov responded by organizing a symposium commemorating a different anniversary: the publication of *Ars Conjectandi* 200 years before.

Markov's strongest vitriol was reserved for another mathematician, Pavel Nekrasov, whose work Markov described as "an abuse of mathematics." Nekrasov was on the faculty of Moscow University, which was then a stronghold of the Russian Orthodox Church. Nekrasov had begun his schooling at a theological seminary before turning to mathematics, and apparently he believed the two vocations could support each other.

In a paper published in 1902 Nekrasov injected the law of large numbers into the centuries-old theological debate about free will versus predestination. His argument went something like this: Voluntary acts—expressions of free will—are like the independent events of probability theory, with no causal links between them. The law of large numbers applies *only* to such independent events. Data gathered by social scientists, such as crime statistics, conform to the law of large numbers. Therefore the underlying acts of individuals must be independent and voluntary.

Markov and Nekrasov stood at opposite poles along many dimensions, a secular republican from Petersburg confronting an ecclesiastical monarchist from Moscow. But when Markov launched his attack on Nekrasov, he did not dwell on factional or ideological differences. He zeroed in on a mathematical error. Nekrasov assumed that the law of large numbers *requires* the principle of independence. Although this notion had been a commonplace of probability theory since the time of Jacob Bernoulli, Markov set out to show that the assumption is unnecessary. The law of large numbers applies perfectly well to systems of dependent variables if they meet certain criteria.

Counting Vowels and Consonants

Markov first addressed the issue of dependent variables and the law of large numbers in 1906. He began with a simple case—a system with just two states. If the states are labeled a and b, there are four possible transitions: $a \to a$, $a \to b$, $b \to b$, $b \to a$. On the assumption that all four transition probabilities are greater than 0 and less than 1, he was able to prove that as the system evolves over time, the frequency of each state converges to a fixed average value. Over the next few years Markov extended and generalized the proof, showing that it applies to a broad class of models.

This series of results achieved at least one of Markov's goals. It forced Nekrasov to retreat from his claim that the law of large numbers implies free will. But the wider world of mathematics did not take much notice. One thing lacking was any hint of how these ideas might be applied to practical events. Markov was proudly aloof from such matters. He wrote to Chuprov, "I am concerned only with questions of pure analysis . . . . I refer to the question of the applicability of probability theory with indifference."

He was too young to have been blighted
by the cold world's corrupt finesse;
his soul still blossomed out, and lighted
at a friend's word, a girl's caress.
In heart's affairs, a sweet beginner,
he fed on hope's deceptive dinner;
the world's éclat, its thunder-roll,
still captivated his young soul.
He sweetened up with fancy's icing
the uncertainties within his heart;
for him, the objective on life's chart
was still mysterious and enticing—
something to rack his brains about,
suspecting wonders would come out.

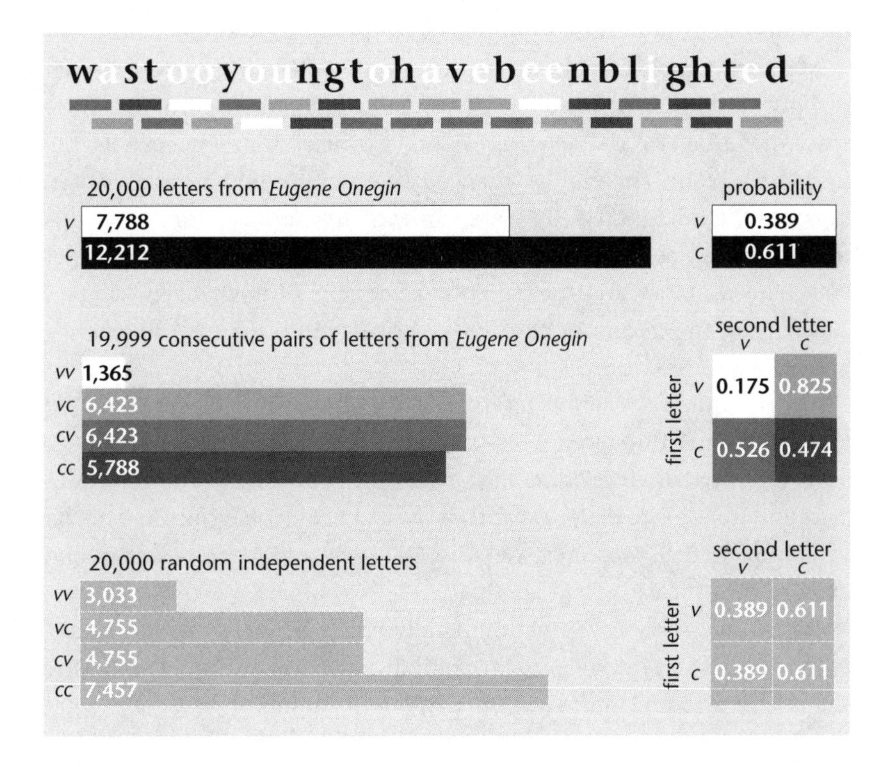

Figure 9.4 Part of Markov's experiment on the statistics of language is repeated with an English translation of *Eugene Onegin* (by Charles H. Johnston) A single stanza (canto 2, verse 7) shows Pushkin's rhyme scheme and meter. For the statistical analysis, punctuation and word spaces are removed, as shown for one phrase in the lower panel. Vowels and consonants are distinguished by white and black type; the four possible pairings of vowels and consonants are marked by shaded underlining. The bar graphs record the tallies of letters and letter pairs, and the 2×2 matrices show probabilities for one kind of letter to follow another. The *Onegin* text exhibits a strong bias in favor of alternation. At the bottom, random pairs drawn from the same set of letters have a different distribution.

By 1913, however, Markov had apparently had a change of heart. His paper on *Onegin* was certainly a work of applied probability theory. It made a lasting impression, perhaps in part because of the novelty of applying mathematics to poetry. Perhaps, too, because the poem he chose is a treasured one, which Russian schoolchildren recite from memory.

From a linguistic point of view, Markov's analysis was at a very superficial level. It did not address the meter or rhyme or meaning of Pushkin's verse. It treated the text as a mere stream of letters. Simplifying further still, the letters were lumped into just two classes, vowels and consonants.

Markov's sample comprised the first 20,000 letters of the poem, which is about one-eighth of the total. He eliminated all punctuation and white space, jamming the characters into one long, unbroken sequence. In the first phase of his analysis he arranged the text in 200 blocks of 10×10 characters, then counted the vowels in each row and column. From this tabulation he was able to calculate both the mean number of vowels per 100-character block and the variance, a measure of how widely samples depart from the mean. Along the way he tallied up the total number of vowels (8,638) and consonants (11,362).

In a second phase Markov returned to the unbroken sequence of 20,000 letters, combing through it to classify pairs of successive letters according to their pattern of vowels and consonants. He counted 1,104 vowel-vowel pairs and was able to deduce that there were 3,827 double consonants; the remaining 15,069 pairs must consist of a vowel and a consonant in one order or the other.

With these numbers in hand, Markov could estimate to what extent Pushkin's text violates the principle of independence. The probability that a randomly chosen letter is a vowel is 8,638/20,000, or about 0.43. If adjacent letters were independent, then the probability of two vowels in succession would be $(0.43)^2$, or about 0.19. A sample of 19,999 pairs would be expected to have 3,731 double vowels, more than three times the actual number. Thus we have strong evidence that the letter probabilities are *not* independent; there is an exaggerated tendency for vowels and consonants to alternate. (Given the phonetic structure of human language, this finding is not a surprise.)

Markov did all his counting and calculating with pencil and paper. Out of curiosity, I tried repeating some of his work with an English translation of *Onegin*. Constructing 10×10 tables on squared paper was tedious but not difficult. Circling double vowels on a printout of the text seemed to go quickly—ten stanzas in half an hour—but it turned out I had missed 62 of 248 vowel-vowel pairs. Markov was probably faster and more accurate than I am; even so, he must have spent several days on these labors. He later undertook a similar analysis of 100,000 characters of a memoir by another Russian writer, Sergei Aksakov.

A computer reduces the textual analysis to triviality, finding all double vowels in 4 milliseconds. The result of such an analysis, shown in figure 9.4, suggests that written English is rather vowel-poor (or consonant-rich) compared with Russian, and yet the structure of the transition matrix is the same. The probability of encountering a vowel depends strongly on whether the preceding letter is a vowel or a consonant, with a bias toward alternation.

The Bootless Academician

Markov's *Onegin* paper has been widely discussed and cited but not widely read outside of the Russian-speaking world. Morris Halle, a linguist at MIT, made an English translation in 1955 at the request of colleagues who were then interested in statistical approaches to language. But Halle's translation was never published; it survives only in mimeograph form in a few libraries. The first widely available English translation, created by the German scholar David Link and several colleagues, was published in 2006.

Link has also written a commentary on Markov's "mathematization of writing" and an account of how the *Onegin* paper came to be known outside of Russia. (A crucial figure in the chain of transmission was George Pólya, a Hungarian mathematician whose well-known work on random walks is closely related to Markov chains.) The statisticians Oscar Sheynin and Eugene Seneta have also written about Markov and his milieu. Because I read no Russian, I have relied heavily on these sources.

In the accounts of Link, Seneta, and Sheynin we find the dénouement of the Markov-Nekrasov conflict. Not surprisingly, the royalist Nekrasov had a hard time hanging on to his position after the 1917 Bolshevik revolution. He died in 1924, and his work fell into obscurity.

Markov, as an anti-tsarist, was looked upon more favorably by the new regime, but an anecdote about his later years suggests he remained a malcontent to the end. In 1921 he complained to the Academy that he could not attend meetings because he lacked suitable footwear. The matter was referred to a committee. In a sign of how thoroughly Russian life had been turned upside down, the chairman was none other than Academician Maxim Gorky. A pair of boots was found for Comrade Markov, but he said they didn't fit and were "stupidly stitched." He continued to keep his distance from the Academy and died in 1922.

The Drivel Generator

For Markov, extending the law of large numbers to interdependent samples was the main point of his inquiry. He bequeathed us a proof that a Markov chain must eventually settle down to some definite, stable configuration corresponding to the long-term average behavior of the system.

In the century since 1913, Markov chains have become a major mathematical industry, but the emphasis has shifted away from the questions that most interested Markov himself. In a practical computational setting, it's not enough to know that a system will *eventually* converge to a stable value; one needs to know how long it will take. With the recent vogue for huge Markov systems, even estimating the convergence time is impractical; the best that can be expected is an estimate of the error introduced by prematurely terminating a simulation process.

I conclude this essay with a more personal story about my own introduction to Markov chains. In 1983 I wrote a "Computer Recreations" column for *Scientific American* subtitled "A progress report on the fine art of turning literature into drivel." I was exploring algorithms that exploit the statistical structure of language to generate random text in the manner of a particular author. (Some specimens based on *Eugene Onegin* appear in figure 9.5.)

One version of the drivel algorithm builds a transition matrix whose rows are labeled by sequences of k letters and whose columns define the probabilities of various letters that can follow each k-character sequence. Given an initial k-letter seed sequence, the program uses the matrix and a random number generator to choose the next character of the synthesized text. Then the leftmost letter of the seed is dropped, the newly chosen character is appended to the right side, and the whole procedure repeats. For values of k larger than 2 or 3 the matrix becomes impractically large, but there are tricks for solving this problem (one of which eliminates the matrix altogether).

Shortly after my article appeared, I met Sergei Kapitsa, the son of the Nobel laureate Pyotr Kapitsa and the editor of the Russian language edition of *Scientific American*. Kapitsa told me that my algorithms for generating random text all derived from the work of A. A. Markov, decades earlier. I expressed a certain skepticism. Maybe Markov invented the underlying mathematics, but did he apply those ideas to linguistic processes? Then Kapitsa told me about Markov's *Onegin* paper.

In a later issue of the magazine I published a contrite addendum about Markov. I had to write it without ever having read a word of Markov's work, and I went overboard a little, saying that Markov "asks to what extent

First order

Theg sheso pa lyiklg ut. cout Scrpauscricre cobaives wingervet Ners, whe ilened te o wn taulie wom uld atimorerteansouroocono weveiknt hef ia ngry'sif farll t mmat and, tr iscond frnid riliofr th Gureckpeag

Third order

At oness, and no fall makestic to us, infessed Russion-bently our then a man thous always, and toops in he roguestill shoed to dispric! Is Olga's up. Italked fore declaimsel the Juan's conven night toget nothem,

Fifth order

Meanwhile with jealousy bench, and so it was his time. But she trick. Let message we visits at dared here bored my sweet, who sets no inclination, and Homer, so prose, weight, my goods and envy and kin.

Seventh order

My sorrow her breast, over the dumb torment of her veil, with our poor head is stooping. But now Aurora's crimson finger, your christening glow. Farewell. Evgeny loved one, honoured fate by calmly, not yet seeking?

Figure 9.5 Random text generated by a Markov chain has statistical properties matching those of *Eugene Onegin* (in the Johnston translation). In a kth-order random text Markov model the states are sequences of k characters, and transitions are defined for all characters that can follow each of these sequences. The probabilities of the transitions are weighted according to frequency. For example, the three-letter sequence *eau* appears 27 times in the text of the poem; in 18 cases it is followed by a *t* (as in *beauty*), in four cases by an *x* (as in *Bordeaux*), and there are four other possible sequels. Thus if the current state of the Markov chain is *eau*, the next state is *aut* with probability $^{18}/_{27} = 0.667$. The first-order model, in which each character's probability depends on a single preceding character, produces gibberish. Text from the third-order Markov chain is mostly pronounceable; by fifth order it consists almost entirely of genuine words; at seventh order whole phrases are plucked from the original text.

Pushkin's poem remains Pushkin's when the letters are scrambled." Thirty years later, I hope this essay will restore the balance. Sadly, though, I am too late to share it with Kapitsa. He died in 2012 at age 84.

I have never quite outgrown my fondness for computer-generated drivel. The program I wrote in 1983 was implemented in Microsoft Basic on one of the earliest IBM PCs. I no longer have the hardware or software to run such programs, so I decided to write a new one. Doing it in Javascript allowed me to make it readily available on the web. Thus you too can now turn fine literature into incoherent nonsense. The program is at http://bit-player.org/wp-content/extras/drivel/drivel.html.

10

Playing Ball in the *n*th Dimension

The area enclosed by a circle is πr^2. The volume inside a sphere is $\frac{4}{3}\pi r^3$. These are formulas I learned too early in life. Having committed them to memory as a schoolchild, I ceased to ask questions about their origin or meaning. In particular, it never occurred to me to wonder how the two formulas are related, or whether they could be extended beyond the familiar world of two- and three-dimensional objects to the geometry of higher-dimensional spaces. What's the volume bounded by a four-dimensional sphere? Is there some master formula that gives the measure of a round object in n dimensions?

Fifty-some years after my first exposure to the formulas for area and volume, I have finally had occasion to look into these broader questions. Finding the master formula for n-dimensional volumes was easy; a few minutes with Google and Wikipedia was all it took. But I've had many a brow-furrowing moment since then trying to make sense of what the formula is telling me. The relation between volume and dimension is not at all what I expected; indeed, it's one of the zaniest things I've ever come upon in mathematics. I'm appalled to realize that I have passed so much of my life in ignorance of this curious phenomenon. I write about it here in case anyone else also missed school on the day the class learned n-dimensional geometry.

Lost in Space

In those childhood years when I was memorizing volume formulas, I also played a lot of ball games. Often the game was delayed when we lost the ball in the weeds beyond right field. I didn't know it then, but we were lucky we played on a two-dimensional field. If we had lost our ball in a space of many dimensions, we might still be looking for it.

The mathematician Richard Bellman labeled this effect "the curse of dimensionality." As the number of spatial dimensions goes up, finding things or measuring their size and shape gets harder. Bellman's curse is illustrated by the ball-in-a-box phenomenon. Put an n-dimensional ball in an n-dimensional cube just large enough to hold it. As n increases, the fraction of the cube's volume occupied by the ball falls dramatically (see figure 10.1). In a space with 100 dimensions, the ball all but vanishes; the occupied fraction is only about $1/10^{70}$. This is far smaller than the volume of an atom in relation to the volume of the Earth.

Since we live in a universe with just three spatial dimensions, Bellman's curse might seem to be of purely philosophical interest, but that's not the case at all. Many computational tasks are carried out in a high-dimensional setting. One common procedure estimates the volume of an object by picking points at random and counting how many lie inside the object. For the 100-dimensional ball in a 100-dimensional cube, you could select a trillion points at random from the interior of the cube and still have almost no chance of finding even one point that is also inside the ball.

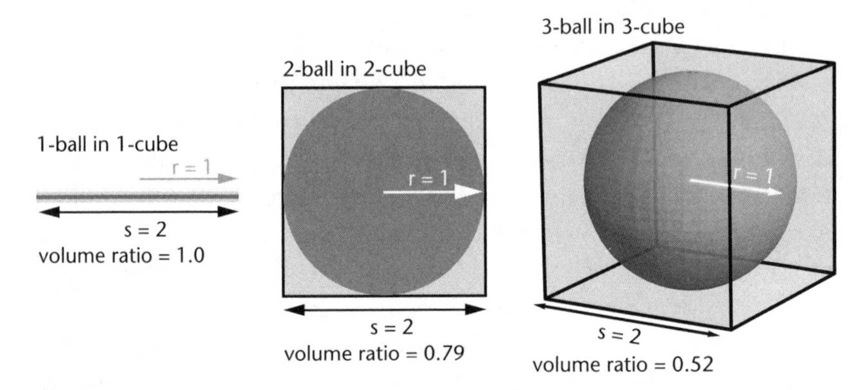

Figure 10.1 Balls in boxes offer a model for studying geometry across spatial dimensions. The boxes are cubes with sides of length 2, which makes them just large enough to accommodate a ball of radius 1. In one dimension *(left)* the ball and the cube have the same shape: a line segment of length 2. In two dimensions *(middle)* and three dimensions *(right)* the balls are more recognizably round. As dimension increases, the ball fills a smaller and smaller fraction of the cube's volume. In three dimensions the filled fraction is about half; in 100-dimensional space *(not shown)*, the ball has all but vanished. The occupied fraction is only about $1/10^{70}$.

Before going further with this adventure, it's worth pausing to consider just what we mean by a ball in n-dimensional space. A three-dimensional ball, or 3-ball, is the kind we throw and catch. It has a center and a radius, r. Mathematically, the ball is the set of all points whose distance from the center is no greater than r. This definition works just as well in two dimensions, where a 2-ball is a disk—a circle and its interior. In one dimension a 1-ball turns out to be simply a line segment of length $2r$. It doesn't look particularly round, but it still satisfies the definition; it includes all points within distance r of the center point.

Going in the other direction, imagination fails me when I try to visualize a ball in four or more dimensions. Nevertheless, the definition continues to provide a recipe for constructing n-balls. What's not so clear is how to calculate the size of those objects.

The Master Formula

Consider an n-ball of radius 1 (a unit ball). The ball will just fit inside an n-cube with sides of length $s = 2$. The surface of the ball kisses the center of each face of the cube. In this configuration, what fraction of the cubic volume is filled by the ball?

The question is answered easily in the low-dimensional spaces we all call home. A 1-ball with $r = 1$ and a 1-cube with $s = 2$ are actually the same object—a line segment of length 2. Thus in one dimension the ball completely fills the cube; the volume ratio is 1.0.

In two dimensions, a 2-ball inside a 2-cube is a disk inscribed in a square, and so this problem can be solved with one of my childhood formulas. With $r = 1$, the area πr^2 is simply π, whereas the area of the square, s^2, is 4; the ratio of these quantities is about 0.79.

In three dimensions, the unit ball's volume is $\frac{4}{3}\pi$, whereas the cube has a volume of $2^3 = 8$; this works out to a ratio of approximately 0.52.

On the basis of these three data points, it appears that the ball fills a smaller and smaller fraction of the cube as n increases. There's a simple intuitive argument suggesting that the trend will continue. In two and three dimensions we can see that the ball occupies the middle of the cube but does not reach deeply into the corners. Each time n increases by 1, the number of corners doubles, so we can expect ever more of the cube's volume to migrate into the nooks and crannies near the vertices.

To get beyond this appealing but nonquantitative intuition, I would have to calculate the volume of n-balls and n-cubes for values of n greater than 3. The calculation is easy for the cube. An n-cube with sides of length s has volume s^n. The cube that encloses a unit ball has $s = 2$, so the volume is 2^n, which takes on values of 2, 4, 8, 16, 32, and so on.

But what about the n-ball? As I have already mentioned, my early education failed to equip me with the necessary formula, and so I turned to the web. In two or three clicks I had before me a Wikipedia page titled "Volume of an n-ball" (https://en.wikipedia.org/wiki/Volume_of_an_n-ball). Near the top of the page was the formula I sought:

$$V(n,r) = \frac{\pi^{\frac{n}{2}} r^n}{\Gamma(\frac{n}{2}+1)}.$$

The equation defines the volume V of a ball with radius r in a space of n dimensions. Later I would look into where the formula came from, both historically and mathematically, but my first impulse at that moment was merely to plug in some numbers for r and n, to see what would come out.

The numerator of the formula is easy enough to evaluate, with π raised to the power of $n/2$ and r raised to the nth power. The expression in the denominator is a little less obvious. It is a gamma function (the symbol Γ is the Greek capital letter gamma). The gamma function is an elaboration on the idea of a factorial. The factorial of a positive integer n, written $n!$, is the product of all the integers from 1 through n. For example, $5! = 1 \times 2 \times 3 \times 4 \times 5 = 120$. The gamma function is very similar when it is applied to an integer; $\Gamma(n)$ is the product of the positive integers less than n, so that $\Gamma(5) = 1 \times 2 \times 3 \times 4 = 24$. Thus $n! = \Gamma(n+1)$. But the gamma function is not just a shifted version of the factorial. The factorial is defined *only* for positive integers, but the gamma function has a value for any positive real number. In particular, $\Gamma(1/2)$ is equal to $\sqrt{\pi}$, and for all odd n, $\Gamma(n/2)$ includes a factor of $\sqrt{\pi}$.

The Incredible Shrinking *n*-Ball

When I discovered the n-ball formula, I wrote a hasty one-line program in Mathematica and began tabulating the volume of a unit ball in various dimensions. I had definite expectations about the outcome. I believed that the volume of the unit ball would increase steadily with n, though at a lower rate than the volume of the enclosing cube, thereby confirming

Bellman's curse of dimensionality. Here are the first few results returned by the program:

n	$V(n,1)$	
1	2	
2	π	≈ 3.1416
3	$\frac{4}{3}\pi$	≈ 4.1888
4	$\frac{1}{2}\pi^2$	≈ 4.9348
5	$\frac{8}{15}\pi^2$	≈ 5.2638

I noted immediately that the values for one, two, and three dimensions agreed with the results I already knew (a comforting confirmation). I also observed that the volume was slowly increasing with n, as I had expected. But then I looked at the continuation of the table:

n	$V(n,1)$	
1	2	
2	π	≈ 3.1416
3	$\frac{4}{3}\pi$	≈ 4.1888
4	$\frac{1}{2}\pi^2$	≈ 4.9348
5	$\frac{8}{15}\pi^2$	≈ 5.2638
6	$\frac{1}{6}\pi^3$	≈ 5.1677
7	$\frac{16}{105}\pi^3$	≈ 4.7248
8	$\frac{1}{24}\pi^4$	≈ 4.0587
9	$\frac{32}{945}\pi^4$	≈ 3.2985
10	$\frac{1}{120}\pi^5$	≈ 2.5502

Beyond the fifth dimension, the volume of a unit n-ball *decreases* as n increases! I tried a few larger values of n, finding that $V(20,1)$ is about 0.0258, and $V(100,1)$ is in the neighborhood of 10^{-40}. It looks like the n-ball dwindles away to nothing as n approaches infinity.

Doubly Cursed

I had thought that I understood Bellman's curse: Both the n-ball and the n-cube grow along with n, but the cube expands faster. In fact, the curse is far more damning. At the same time the cube inflates exponentially, the ball shrinks to insignificance. In a space of 100 dimensions, the cube has grown to a volume of 1.3×10^{30} while the ball has shrunk to 2.4×10^{-40}. The fraction of the cubic volume filled by the ball is 1.9×10^{-70}.

What makes this disappearing act so extraordinary is that the ball in question is still the largest one that could possibly be stuffed into the cube. We are not talking about a pea rattling around loose inside a refrigerator carton. The ball's diameter is still equal to the side length of the cube. The surface of the ball touches every face of the cube. The fit is snug; if the ball were made even a smidgen larger, it would bulge out of the cube on all sides. Nevertheless, in terms of volume measure, the ball is nearly crushed out of existence, like a black hole collapsing under its own mass.

How can we make sense of this seeming paradox? One way of understanding it is to acknowledge that the ball fills the middle of the cube, but the cube doesn't have much of a middle; almost all its volume is away from the center, huddling in the corners. As noted, the ball touches the enclosing cube at the center of each face, but it does not reach into the corners. A 100-cube has just 200 faces, but it has 2^{100} corners.

Another approach to understanding the collapse of the n-ball is to imagine poking skewers through the cube along various diameters. (A diameter is any straight line that passes through the center point.) The shortest diameters run from the center of a face to the center of the opposite face. For the cube enclosing a unit ball, the length of this shortest diameter is 2, which is both the side length of the cube and the diameter of the ball. Thus a skewer on the shortest diameter lies inside the ball throughout its length.

The longest diameters of the cube extend from a corner through the center point to the opposite corner. For an n-cube with side length $s = 2$, the length of this diameter is $2\sqrt{n}$. Thus in the 100-cube surrounding a unit ball, the longest diameter has length 20; only 10 percent of this length lies within the ball. Moreover, there are just 100 of the shortest diameters, but there are 2^{99} of the longest ones.

Here is still another mind-bending trick with balls and boxes to suggest just how weird space becomes in higher dimensions. I learned of it from Barry Cipra, who published a description in volume 1 of *What's Happening in the Mathematical Sciences* (1991). On the plane, a square with sides of length 4 will accommodate four unit disks in a 2×2 array, with room for a smaller disk in the middle (see figure 10.2). The radius of the central disk is $\sqrt{2} - 1$. In three dimensions the equivalent 3-cube fits eight unit balls, plus a smaller ninth ball in the middle, whose radius is $\sqrt{3} - 1$. In the general case

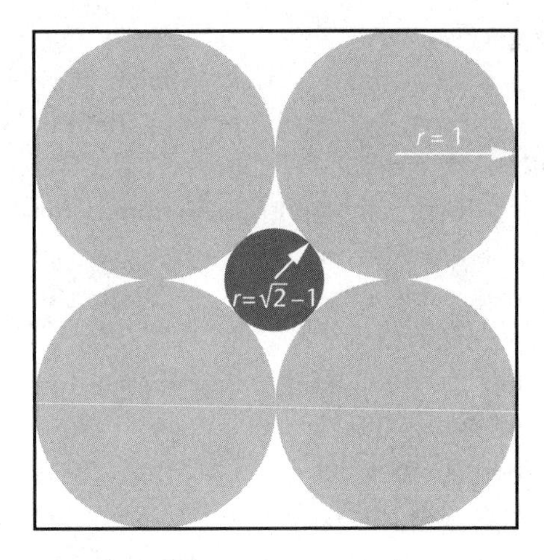

Figure 10.2 Four 2-balls squeezed into a 2-cube leave room for a smaller ball between them. In higher dimensions the "smaller" ball grows larger than those surrounding it. Beyond dimension 9 it bursts through the sides of the confining cube.

of n dimensions, the box has room for 2^n unit n-balls in a rectilinear array, with one additional ball in the vacant central space, and the central ball has a radius of $\sqrt{n}-1$. Look what happens when n reaches 9. The "smaller" central ball now has a radius of 2, which makes it twice the size of the 512 surrounding balls. Furthermore, the central ball has expanded to reach the sides of the bounding box, and will burst through the walls with any further increase in dimension.

What's So Special About the 5-Ball?

I was taken by surprise when I learned that the volume of a unit n-ball goes to zero as n goes to infinity; I had expected the opposite. But something else surprised me even more—the fact that the volume function is not monotonic. Either a steady increase or a steady decrease seemed more plausible than having the volume grow for a while, then reach a peak at some finite value of n, and thereafter decline. This behavior singles out a particular dimension for special attention. What is it about a space of five dimensions space that allows a unit 5-ball to spread out more expansively than any other n-ball?

I can offer an answer, although it doesn't really explain much. The answer is that everything depends on the value of π. Because π is a little more than 3, the volume peak comes at five dimensions; if π were equal to 17, say, the unit ball with maximum volume would be found in a space with 33 dimensions.

To see how π comes to have this role, we'll have to return to the formula for n-ball volume. We can get a rough sense of the function's behavior from a simplified version of the formula. If we are interested only in the unit ball, then r is always equal to 1, and the r^n term can be ignored. That leaves a power of π in the numerator and a gamma function in the denominator. If we consider only even values of n, so that $n/2$ is always an integer, we can replace the gamma function with a factorial. For brevity, let $m = n/2$; then all that remains of the formula is this ratio: $\pi^m/m!$.

The simplified formula says that the n-ball volume is determined by a race between π^m and $m!$. Initially, for the smallest values of m, π^m sprints ahead; for example, at $m = 2$ we have $\pi^2 \approx 10$, which is greater than $2! = 2$. In the long run, however, $m!$ will surely win this race. Both π^m and $m!$ are products of m factors, but in π^m the factors are all equal to π, whereas in $m!$ they range from 1 up to m. Numerically, $m!$ first exceeds π^m when $m = 7$, and thereafter the factorial grows very much larger.

This simplified analysis accounts for the major features of the volume curve, at least in a qualitative way. The volume of a unit ball has to go to zero in infinite-dimensional space because zero is the limit of the ratio

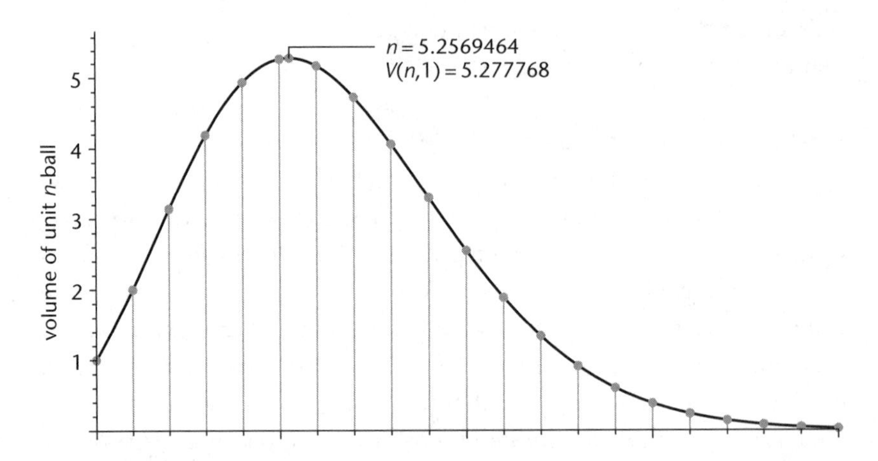

Figure 10.3 The volume of a unit ball in n dimensions reveals an intriguing spectrum of variations. Up to dimension 5, the ball's volume increases with each increment to n; then the volume starts diminishing again, and ultimately goes to zero as n goes to infinity. If the dimension of space is considered a continuous variable, the peak volume comes at $n = 5.2569464$.

$\pi^m/m!$. In the lowest dimensions, on the other hand, the ratio is increasing with m. And if it's going uphill for small m and downhill for large m, there must be some intermediate value where the function reaches a maximum.

To get a quantitative fix on the location of this maximum, we must return to the formula in its original form and consider odd as well as even numbers of dimensions. Indeed, we can take a step beyond mere integer dimensions. Because the gamma function is defined for all real numbers, we can treat dimension as a continuous variable and ask with finer resolution where the maximum volume occurs. A numerical solution to this calculus problem—found with further help from Mathematica—shows a peak in the volume curve at $n \approx 5.2569464$; at this point the unit ball has a volume of 5.2777680 (see figure 10.3).

With a closely related formula, we can also calculate the surface area of an n-ball. Like the volume, this quantity reaches a peak and then falls away to zero. The maximum is at $n \approx 7.2569464$, two dimensions larger than the volume peak.

The Dimensions of the Problem

The arithmetic behind all these results is straightforward; attaching meaning to the numbers is not so easy. In particular, I can see numerically—by comparing powers of π with factorials—why the unit ball's volume reaches a maximum at $n = 5$. But I have no geometric intuition about five-dimensional space that would explain this fact.

The results on noninteger dimensions are quite otherworldly. The notion of fractional dimensions is familiar enough, but it is generally applied to objects, not to spaces. For example, the Sierpiński triangle, with its endlessly nested holes within holes, is assigned a dimension of 1.585, but the triangle is still drawn on a plane of dimension 2. What would it mean to construct a space with 5.2569464 mutually perpendicular coordinate axes? I can't imagine—and that's not just a figure of speech.

Another troubling question is whether it really makes sense to compare volumes across dimensions. Each dimension requires its own units of measure, and so the relative magnitudes of the numbers attached to those units don't mean much. Is a disk of area 10 square centimeters larger or smaller than a ball of volume 5 cubic centimeters? We can't answer; it's like comparing apples and orange juice.

One way to eliminate all units is to look only at ratios of volumes. In each dimension volume is measured in terms of a standard volume *in that dimension.* The obvious standard is the unit cube (sometimes called the "measure polytope"), which has a volume of 1 in all dimensions. Starting at $n = 1$, the unit ball is larger than the unit cube, and the ball-to-cube ratio gets still larger through $n = 5$; then the trend reverses, and eventually the ball is much smaller than the unit cube. This changing ratio of ball volume to cube volume is the phenomenon to be explained.

When an earlier version of this essay was published in *American Scientist,* a few readers expressed continuing skepticism even of a formulation based on ratios, and suggested there is nothing really surprising about the sequence of n-ball volumes. Perhaps when I am truly at home on 100-dimensional space, I, too, will see these numbers as natural or even inevitable, but I'm not there yet.

Slicing the Onion

The volume formulas I learned as a child were incantations to be memorized rather than understood. I would like to do better now. Although I cannot give a full derivation of the n-ball formula—for lack of both space and mathematical acumen—perhaps the following remarks will shed some light.

The key idea is that an n-ball has within it an infinity of $(n-1)$-balls. For example, a series of parallel slices through the body of an onion turns a 3-ball into a stack of 2-balls. Another set of cuts, perpendicular to the first series, reduces each disk-like slice to a collection of 1-balls—linear ribbons of onion. If you go on to dice the ribbons, you have a heap of 0-balls. (With real onions and knives these operations only approximate the forms of true n-balls, but the methods work perfectly in the mathematical kitchen.)

This decomposition suggests a recursive algorithm for computing the volume of an n-ball: Slice it into many $(n-1)$-balls and sum up the volumes of the slices. How do you compute the volumes of the slices? Apply the same method, cutting the $(n-1)$-balls into $(n-2)$-balls. Eventually the recursion bottoms out at $n = 1$ or $n = 0$, where the answers are known. (The volume of a 1-ball is $2r$; the 0-ball is assigned a volume of 1.) Letting the thickness of the slices go to zero turns the sum into an integral and leads to an exact result.

In practice, it's convenient to use a slightly different recursion with a step size of 2. That is, the volume of an n-ball is computed from that of an $(n-2)$-ball. The specific rule is, given the volume of an $(n-2)$-ball, multiply by $2\pi r^2/n$ to get the volume of the corresponding n-ball. (Showing *why* the multiplicative factor takes this particular form is the hard part of the derivation, which I am going to gingerly avoid; it requires an exercise in multivariable calculus that lies beyond my abilities.)

The procedure is easy to express in the form of a computer program:

```
function V(n, r)
    if n = 0 then return 1
    elseif n = 1 then return 2r
    else return 2πr2/n × V(n−2, r)
```

For even n, the sequence of operations carried out by this program amounts to

$$1 \times \frac{2\pi r^2}{2} \times \frac{2\pi r^2}{4} \times \frac{2\pi r^2}{6} \times \cdots \times \frac{2\pi r^2}{n}.$$

For odd n, the result is instead the product of these terms:

$$2r \times \frac{2\pi r^2}{3} \times \frac{2\pi r^2}{5} \times \frac{2\pi r^2}{7} \times \cdots \times \frac{2\pi r^2}{n}.$$

For all integer values of n the program yields the same output as the formula based on the gamma function.

Who Done It?

A question I cannot answer with certainty is who first wrote down the n-ball formula. I have paddled up a long river of references, but I'm not sure I have reached the true source.

My journey began with the number 5.2569464. I entered the digits into the *On-Line Encyclopedia of Integer Sequences,* the vast compendium of number lore created by Neil J. A. Sloane. I found what I was looking for in sequence A074455. A reference there directed me to *Sphere Packings, Lattices, and Groups,* by John Horton Conway and Sloane. That book in turn cited *An Introduction to the Geometry of N Dimensions,* by Duncan Sommerville, published in 1929. The Sommerville book devotes a few pages to the n-ball formula and has a table of values for dimensions 1 through 7, but it says little about origins. However, further sleuthing in library catalogs revealed that Sommerville, a Scottish mathematician who emigrated to New Zealand in 1915, also published a bibliography of non-Euclidean and n-dimensional geometry.

The bibliography lists five works on "hypersphere volume and surface"; the earliest is a problem and solution published in 1866 by William Kingdon Clifford, a brilliant English geometer who died young. Clifford's derivation of the formula is clearly original work, but it was not the first.

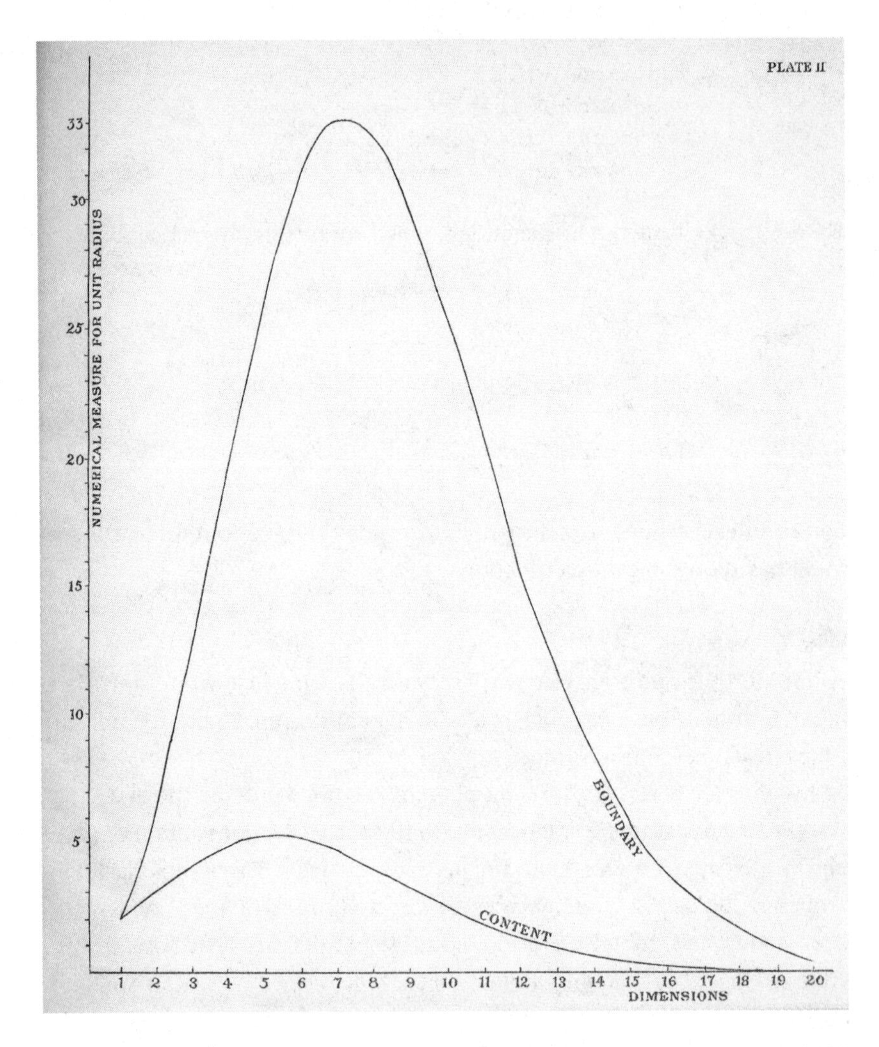

Figure 10.4 This graph of n-ball volume as a function of dimension was plotted 120 years ago by Paul Renno Heyl, who was then a graduate student at the University of Pennsylvania. The volume graph is the lower curve, labeled "content." The upper curve gives the ball's surface area, for which Heyl used the term "boundary." The illustration is from Heyl's 1897 thesis, "Properties of the locus r = constant in space of n dimensions."

Elsewhere Sommerville mentions the Swiss mathematician Ludwig Schläfli as a pioneer of n-dimensional geometry. Schläfli's treatise on the subject, written in the early 1850s, was not published in full until 1901, but an excerpt translated into English by Arthur Cayley appeared in 1858. The first paragraph of that excerpt gives the volume formula for an n-ball, commenting that it was determined "long ago." An asterisk leads to a footnote citing papers published in 1839 and 1841 by the Belgian mathematician Eugène Catalan.

Looking up Catalan's articles, I found that neither of them gives the correct formula in full, although they're close. Catalan deserves partial credit.

Not one of these early works pauses to comment on the implications of the formula—the peak at $n = 5$ or the trend toward zero volume in high dimensions. Of the works mentioned by Sommerville, the only one to make these connections is a thesis by Paul Renno Heyl, published by the University of Pennsylvania in 1897. This looked like a fairly obscure item, but with help from Harvard librarians, the volume was found on a basement shelf. I later discovered that the full text (but not the plates) is available on Google Books. The omission of the plates is unfortunate; one of them shows what may well be the first published graph of the volume of an n-ball as a function of n (see figure 10.4).

Heyl was a graduate student at the time of this work. He went on to a career with the National Bureau of Standards, and he was also a writer on science, philosophy, and religion. (His best-known book was *The Mystery of Evil*.)

In the 1897 thesis Heyl derives formulas for both volume and surface area (which he calls "content" and "boundary"), and gives a lucid account of multidimensional geometry in general. He clearly appreciates the strangeness of the discovery that "in a space of infinite dimension our locus can have no content at all." I will allow Heyl to have the last word on the subject:

> We might be pardoned for supposing that in a space of infinite dimension we should find the Absolute and Unconditioned if anywhere, but we have reached an opposite conclusion. This is the most curious thing I know of in the Wonderland of Higher Space.

11

Quasirandom Ramblings

In the early 1990s Spassimir Paskov, who was then a graduate student at Columbia University, began analyzing an exotic financial instrument called a collateralized mortgage obligation, or CMO, issued by the investment bank Goldman Sachs. The aim was to estimate the current value of the CMO, based on the potential future cash flow from thousands of 30-year mortgages. This task wasn't just a matter of applying the standard formula for compound interest. Many home mortgages are paid off early when the home is sold or refinanced; some loans go into default; interest rates rise and fall. Thus the present value of a 30-year CMO depends on 360 uncertain and interdependent monthly cash flows. The task amounts to evaluating an integral, or measuring a volume, in 360-dimensional space.

There was no hope of finding an exact solution. Paskov and his adviser, Joseph Traub, decided to try a somewhat obscure approximation technique called the quasi–Monte Carlo method. An ordinary Monte Carlo evaluation takes random samples from the set of all possible solutions. The *quasi* variant does a different kind of sampling—not quite random but not quite regular, either. Paskov and Traub found that some of their quasi–Monte Carlo programs worked far better and faster than the traditional technique. Their discovery would allow a banker or investor to assess the value of a CMO with just a few minutes of computation instead of several hours.

It would make a fine story if I could now report that the subsequent period of "irrational exuberance" in the financial markets—the frenzy of trading in complex derivatives, and the sad sequel of crisis, collapse, recession, unemployment—could all be traced back to a mathematical innovation in the evaluation of high-dimensional integrals. But it's just not so; there were other causes of that folly.

However, the work of Paskov and Traub *did* have an effect. It brought a dramatic revival of interest in quasi–Monte Carlo models. Earlier theoretical results had suggested that the quasi–Monte Carlo technique would begin to run out of steam when the number of dimensions exceeded 10 or 20, and certainly long before it reached 360. Thus the success of the CMO experiment was a surprise, which mathematicians have scrambled to explain. A key question is whether the same approach will work for other problems.

The whole affair highlights the curiously ambivalent role of randomness in computing. Algorithms are by nature strictly deterministic, yet many of them seem to benefit from a small admixture of randomness—an opportunity, every now and then, to make a choice by flipping a coin. In practice, however, the random numbers supplied to computer programs are almost never truly random. They are *pseudorandom*—artful fakes, meant to look random and pass statistical tests, but coming from a deterministic source. What's intriguing is that the phony random numbers seem to work perfectly well, at least for most tasks.

Quasirandom numbers take the charade a step further. They don't even make the effort to dress up and *look* random. Yet they, too, seem to be highly effective in many places where randomness is called for. They may even outperform their pseudorandom cousins in certain circumstances.

Integration by Darts

Here's a toy problem to help pin down the distinctions between the *pseudo* and *quasi* varieties of randomness. Suppose you want to estimate the area of an object with a complicated shape, such as a maple leaf. There's a well-known trick for solving this problem with the help of a little randomness. Put the leaf on a board of known area, then throw darts at it randomly, trying *not* to aim. If a total of N darts hit the board, and n of them land within the leaf, then the ratio $n:N$ approximates the ratio of the leaf area to the board area. For convenience we can define the board area as 1, so the estimated leaf area is simply n/N.

Tossing darts at random can be difficult and dangerous. But if you're willing to accept dots in lieu of darts, and if you let the computer take care of sprinkling them at random, the leaf-measuring experiment is easy, and it works remarkably well.

I collected a leaf from a nearby maple, photographed it, and placed the shape within a square field of 1,024 × 1,024 blank pixels (see figure 11.1). Then I wrote a program that tosses random dots (actually pseudorandom dots) at the digitized image. The first time I ran the program, 429 of 1,024 dots fell on

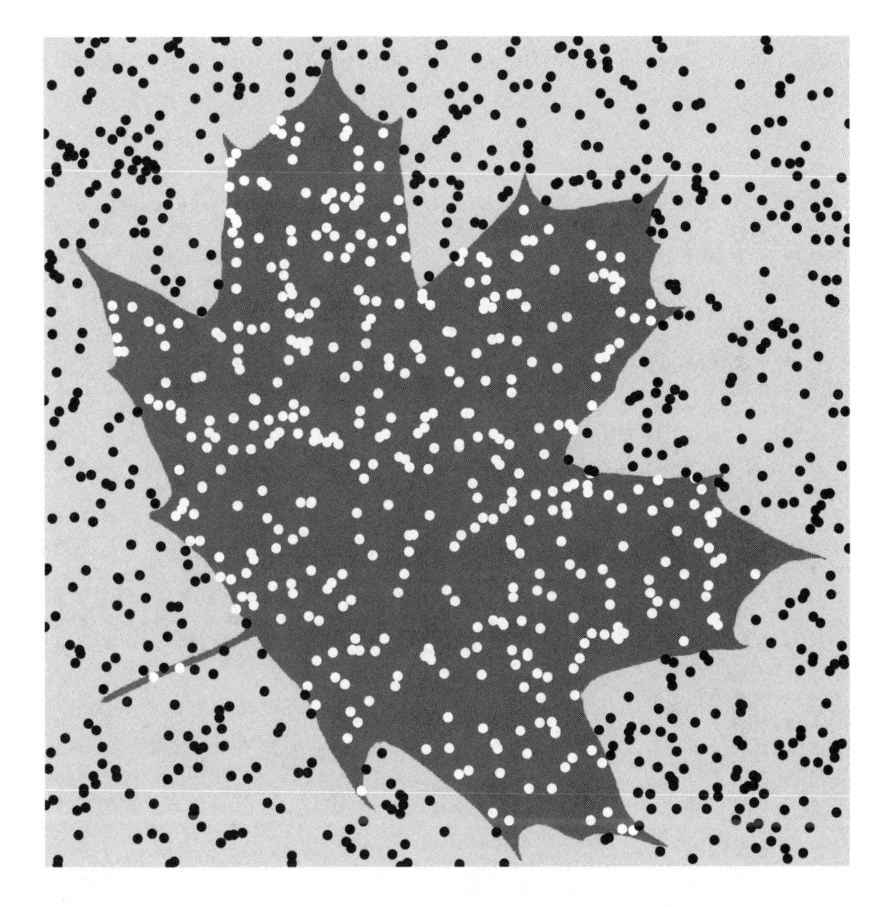

Figure 11.1 Estimating the area of a complex form—in this case a maple leaf—is a classic application of the computational device known as the Monte Carlo method. The image of the leaf is embedded in a square whose area can be taken as 1. The idea is to sprinkle dots throughout the square and count how many hit the leaf *(white dots)* and how many miss *(black dots)*. The fraction of hits approximates the area of the leaf. In this instance 1,024 dots are scattered randomly over the square. When the experiment was repeated 1,000 times, the mean fraction of hits was 0.4183, quite close to the true area of 0.4185.

the leaf, for an area estimate of 0.4189. The actual area, as defined by a count of colored pixels, is 0.4185. The random dots approximation was somewhat better than I expected—a case of random good luck, though nothing out of the ordinary. When I repeated the experiment 1,000 times, the mean estimate of the leaf's area was 0.4183, with a standard deviation of 0.0153.

The basic idea in Monte Carlo studies is to reformulate a mathematical problem, such as calculating the area of a leaf, as a game of chance, where a player's expected winnings are the answer to the problem. In simple games,

Figure 11.2 Randomness is generally taken to be an essential element of the Monte Carlo method, but the leaf-measuring experiment succeeds even when the dots are arranged in a crystalline lattice. Here 1,024 dots in a 32×32 grid yield an area estimate of 0.4209, slightly further from the true value 0.4185.

Figure 11.3 The quasi–Monte Carlo method employs a distinctive sampling pattern that lies somewhere between random and regular. The aim is to ensure a uniform density of dots without long-range order. For the maple leaf the quasirandom area estimate is 0.4141, about 1 percent smaller than the true area.

one can calculate the exact probability of every outcome, so the expected winnings can also be determined exactly. Where such calculations are infeasible, the alternative is to go ahead and play the game, and see how it comes out. This is the strategy of a Monte Carlo simulation. The computer plays the game many times and takes the average result as an estimate of the true expected value.

For the leaf-in-a-square problem, the expected value is the ratio of the leaf area to that of the square. Choosing N random points and counting the number of hits approximates the area ratio, and the approximation gets better as N increases. When N approaches infinity, the measurement becomes exact. This last point is not just an empirical observation but a promise made by a mathematical theorem, namely, the law of large numbers. This is the same principle that guarantees a fair coin will come up heads half the time when the sample is large enough.

The Curse of Dimensionality

Randomness has a conspicuous role in this description of the Monte Carlo method. In particular, the appeal to the law of large numbers requires that the sample points be chosen randomly. And it's easy to see—just by looking at a picture—why random sampling works well: it scatters points everywhere. What's *not* so easy to see is why other kinds of sample point arrangements would not also serve the purpose. After all, one could measure the leaf area by laying down a rectilinear grid of points in the square and counting the hits. I tried this experiment with my leaf image, placing 1,024 points in a 32×32 grid (see figure 11.2). I got an area estimate of 0.4209—not quite as good as my lucky random run, but still within 0.6 percent of the true area.

I also tried measuring the leaf with 1,024 quasirandom sample points, whose arrangement is in some sense intermediate between total chaos and total order. (For an explanation of how the quasirandom pattern is constructed, see the box "A Recipe for Quasirandom Numbers" later in this chapter.) The estimate from counting hits with quasirandom points was 0.4141, giving an error of 1 percent (see figure 11.3).

All three of these procedures give quite respectable results. Does that mean they are all equally powerful? No, I think it means that measuring the area of a leaf in two dimensions is an easy problem.

The task gets much harder in higher dimensions. Understanding why calls for an exercise in multidimensional thinking. Imagine a d-dimensional "cube" with edges of length 1, and a smaller cube inside it, with edge lengths along each dimension equal to $\frac{1}{2}$ (see figure 11.4). When $d = 1$, a cube is just a line segment, and volume is equivalent to length; thus the smaller cube has half the volume of the large one. For $d = 2$, the cube is a square, and volume is area; the small cube has volume $\frac{1}{4}$. The case $d = 3$ corresponds to an ordinary cube, and the volume filled by the small cube is now just $\frac{1}{8}$. The progression continues. By the time we reach dimension

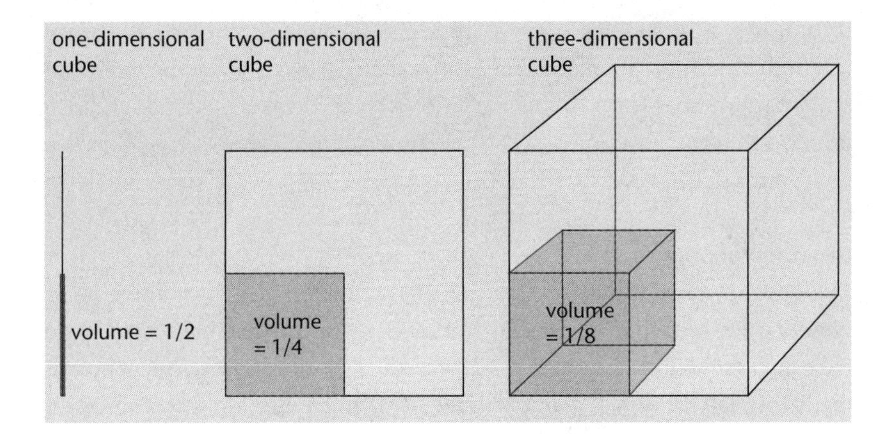

Figure 11.4 Estimating volumes gets progressively harder in higher-dimensional spaces, an effect called "the curse of dimensionality." Suppose a d-dimensional cube with edges of length 1 has a smaller cube inside it with edges of length ½. When $d = 1$ (in which case the "cube" is actually a line segment, and volume is length), the smaller cube fills half the total volume. For $d = 2$ (a square, with volume equivalent to area), the fractional volume is ¼, and for $d = 3$ it is ⅛. The fractional volume shrinks as $1/2^d$, which means that roughly 2^d sampling points are needed to detect the existence of the small cube.

$d = 20$, the smaller cube—still with edge length ½ along each dimension—occupies only about one-millionth of the total volume. The mathematician Richard Bellman called this phenomenon "the curse of dimensionality."

If we want to measure the volume of the one-in-a-million small cube—or even just detect its presence—we need enough sampling points to get at least one sample from the small cube's interior. In other words, when we're counting hits, we need to count at least one. For a 20-dimensional grid pattern, that means we need 1 million points (or, more precisely, $2^{20} = 1{,}048{,}576$). With random sampling, the size requirement is probabilistic and hence a little fuzzy, but the number of points needed if we want to have an expectation of a single hit is again 2^{20}. The analysis for quasirandom sampling comes out the same. Indeed, if you are groping blindly for an object of volume $1/2^d$, it hardly matters how the search pattern is arranged; you will have to look in 2^d places.

If real-world problems were as hard as this one, the situation would be bleak. There would be no hope at all of dealing with a 360-dimension integral. But we know some problems of that scale do yield to Monte Carlo techniques; a reasonable guess is that the solvable problems have some internal structure that speeds the search. Furthermore, the choice of sampling pattern does seem to make a difference, so there is a meaningful distinction to be made among all the gradations of *true, pseudo, quasi,* and *non* randomness.

Random Variations

The concept of randomness in a set of numbers has at least three components. First, randomly chosen numbers are unpredictable; there is no fixed rule governing their selection. Second, the numbers are independent, or uncorrelated; knowing one number will not help you guess another. Finally, random numbers are unbiased, or uniformly distributed; no matter how you slice up the space of possible values, each region can expect to get its fair share.

These concepts provide a useful key for distinguishing between truly random, pseudorandom, quasirandom, and orderly sets. Truly random numbers have all three characteristics: they are unpredictable, uncorrelated, and unbiased. Pseudorandom numbers abandon unpredictability; they are generated by a definite arithmetic rule, and if you know the rule, you can reproduce the entire sequence. But pseudorandom numbers are still uncorrelated and unbiased (at least to a good approximation).

Quasirandom numbers are both predictable and highly correlated. There's a definite rule for generating them, and the patterns they form, although not as rigid as a crystal lattice, nonetheless have a lot of regularity. The one element of randomness that quasirandom numbers preserve is the uniform or equitable distribution. They are spread out as fairly and evenly as possible.

A highly ordered set, such as a cubic lattice, preserves none of the properties of randomness. It's obvious that these points fail the tests of unpredictability and independence, but perhaps it's not so clear that they lack uniform distribution. After all, it's possible to carve up an N-point lattice into N small cubes that each contain exactly one point. But that's not enough to qualify the lattice as fairly and evenly distributed. The ideal of equal distribution demands an arrangement of N points that yields one point per region when the space is divided into *any* set of N identical regions. The rectilinear lattice fails this test when the regions are slices parallel to the axes.

These three aspects of randomness are important in different contexts. In the case of the *other* Monte Carlo—the casino in Monaco—unpredictability is everything. Cryptographic applications of randomness are similar. In these cases, an adversary is trying to detect and exploit any hint of pattern.

Some kinds of computer simulations are very sensitive to correlations between successive random numbers, so independence is important. For the volume estimations under discussion here, however, uniformity of distribution is what matters most. And the quasirandom numbers, by giving up all pretense to unpredictability or lack of correlation, are able to achieve higher uniformity.

A Slight Discrepancy

Uniformity of distribution is measured in terms of its opposite, which is called discrepancy. For points in a two-dimensional square, discrepancy is calculated as follows. Consider all the rectangles with sides parallel to the coordinate axes that could possibly be drawn inside the square. For each such rectangle, count the number of points enclosed, and also calculate the number of points that *would* be enclosed, based on the area of the rectangle, if the distribution were perfectly uniform. The maximum difference between these numbers, taken over all possible rectangles, is a measure of the discrepancy.

Another measure of discrepancy, called star discrepancy, looks only at the subset of axis-parallel rectangles that have one corner anchored at the origin of the unit square (see figure 11.5). Star discrepancy is easier to calculate than general discrepancy. Other measures can be defined, such as schemes based on circles or triangles rather than rectangles. The measures are not all equivalent; results vary depending on the shape. It's interesting to note that the process for measurement of discrepancy has a strong resemblance to the Monte Carlo process itself.

Grids and lattices fare poorly when it comes to discrepancy because the points are arranged in long rows and columns. An infinitely skinny rectangle that should enclose no points at all can take in an entire row. From such worst-case rectangles it follows that the discrepancy of a square lattice with N points is $\sqrt{N}$. Interestingly, it turns out that the discrepancy of a random or pseudorandom lattice is also about $\sqrt{N}$. In other words, in an array of a million random points, there is likely to be at least one rectangle that has either 1,000 points too many or 1,000 too few.

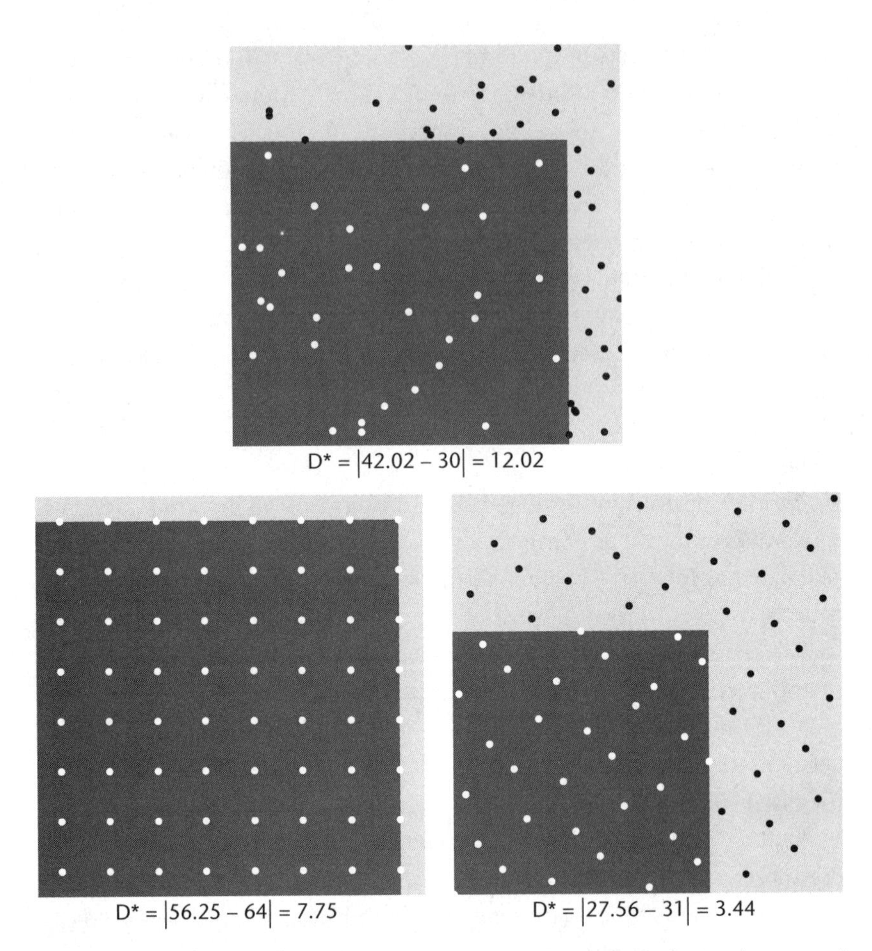

Figure 11.5 Discrepancy measures how far a set of points departs from a uniform spatial distribution. The particular measure illustrated here, called star discrepancy (denoted D*), is defined in terms of rectangles that have one vertex anchored to the lower left corner. For any such rectangle D* is the difference between the expected number of points enclosed and the actual number. For the pattern as a whole D* is given by the worst-case rectangle, the one for which the difference is maximum. Here the star discrepancy is measured for three patterns of 64 points: a pseudorandom array *(top)*, a lattice *(left)*, and a quasirandom array *(right)*. The quasirandom configuration is designed to minimize discrepancy.

Quasirandom patterns are deliberately designed to thwart all opportunities for drawing such high-discrepancy rectangles. For quasirandom points, the discrepancy can be as low as the logarithm of N, which is much smaller than $\sqrt{N}$. For example, at $N = 10^6$, $\sqrt{N} = 1{,}000$ but $\log N = 20$. (I am taking logarithms to the base 2.)

Discrepancy is a key factor in gauging the performance of Monte Carlo and quasi–Monte Carlo simulations. It determines the level of error or statistical imprecision to be expected for a given sample size. For conventional Monte Carlo, with random sampling, the expected error diminishes as $1/\sqrt{N}$. For quasi–Monte Carlo, the corresponding convergence rate is $(\log N)^d/N$, where d is the dimensionality of the space. (Various constants and other details are neglected here, so the comparison is valid only for rates of growth, not for exact values.)

The $1/\sqrt{N}$ convergence rate for random sampling can be painfully slow. Getting one more decimal place of precision requires increasing N by a factor of 100, which means that a one-hour computation suddenly takes four days. The reason for this sluggishness is the clumpiness of the random distribution. Two points that happen to lie close together waste computer effort, because the same region is sampled twice; on the other hand, voids between points leave some areas unsampled and thereby contribute to the error budget.

Compensating for this drawback, random sampling has one very important advantage: the convergence rate does not depend on the dimension of the space. To a first approximation, the program runs as fast in 100 dimensions as in two dimensions. Quasi–Monte Carlo is different in this respect. If only we could ignore the factor of $(\log N)^d$, the performance would be dramatically superior to random sampling. We would have a $1/N$ convergence rate, which calls for just a tenfold increase in effort for each additional decimal place of precision. However, we *can't* ignore the $(\log N)^d$ part. This factor grows exponentially with increasing dimension d. The effect is small in dimension 1 or 2, but eventually it becomes enormous. In ten dimensions, for example, $(\log N)^{10}/N$ remains larger than $1/\sqrt{N}$ until N exceeds 10^{43}.

It was this known slow convergence in high dimensions that led to surprise over the success of the Paskov and Traub financial calculation with $d = 360$. The only plausible explanation is that the "effective dimension" of the problem is actually much lower than 360. In other words, the volume being measured doesn't really extend through all the dimensions of the space. (In the same way, a sheet of paper lives in a three-dimensional space, but not much is lost by pretending it has no thickness.)

A Recipe for Quasirandom Numbers

Numbers that form low-discrepancy patterns were studied long before the quasi–Monte Carlo method was invented. In the 1930s the Dutch mathematician Johannes G. van der Corput asked whether an infinite sequence of distinct numbers could be placed one by one on the unit interval (the segment of the number line from 0 to 1) in such a way that the discrepancy measured at each step would never exceed some finite bound. The answer, proved a decade later by another Dutch mathematician, Tatyana van Aardenne-Ehrenfest, is no; the discrepancy grows without limit. Nevertheless, van der Corput's work gave rise to a whole family of low-discrepancy quasirandom sets and sequences.

The procedure for building a van der Corput sequence is peculiar in that it mixes up numbers (mathematical entities) with numerals (the representation of numbers as lists of digits). The basic idea is to take an integer, reverse its digits, put a decimal point in front of it, and then treat the result as a fraction between 0 and 1. The operations can be carried out in any base. In base 2, for example, the number 100 (equal to decimal 4) becomes 001 and then 0.001, which is the binary representation of the fraction ⅛.

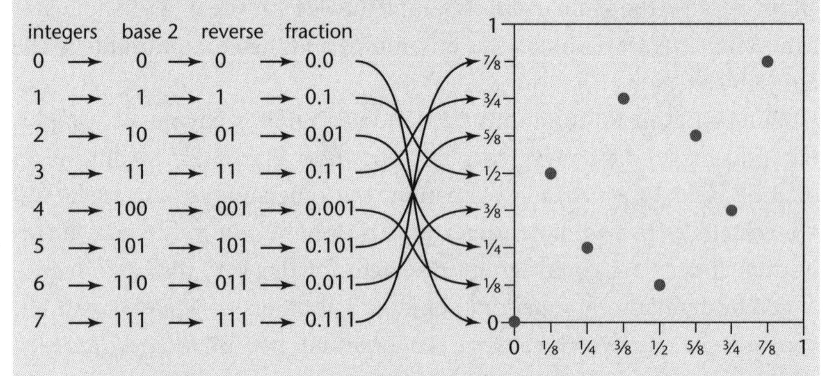

To create a two-dimensional quasirandom pattern with N points, we start with the sequence of integers $i = 0, 1, 2, ..., N-1$. Then for each point the x coordinate is set equal to i/N and the y coordinate is given by the van der Corput digit-reversal process for i. The result for $N = 8$ is shown above. Many other algorithms for low-discrepancy sequences have been developed, but most of them also rely on shuffling the digits of a numeral.

Patterns created by such mechanisms maintain a delicate balance between order and disorder. The spacing between points is reasonably uniform, so there are none of the clumps and voids that raise the discrepancy of a random set. But the uniform distribution has to be achieved without letting the points form rows or other regular structures that would also increase the discrepancy. Balancing these conflicting goals is not hard in two dimensions, but compromise is unavoidable in higher dimensions.

This explanation may sound dismissive. The calculation succeeded not because the tool was more powerful but because the problem was easier than it looked. But note that the effective reduction in dimension works even when we don't know *which* of the 360 dimensions can safely be ignored. That's almost magical.

How commonplace is this phenomenon? Is it just a fluke, or confined to a narrow class of problems? The answer is not yet entirely clear, but a notion called "concentration of measure" offers a reason for optimism. It suggests that the high-dimension world is mostly a rather smooth and flat place, analogous to a high-gravity planet where it costs too much to create jagged alpine landscapes.

Homeopathic Randomness

The Monte Carlo method is not a new idea, and neither is the quasi–Monte Carlo variation. Simulations based on random sampling were attempted more than a century ago by Francis Galton, Lord Kelvin, and others. In those days they worked with true random numbers (and had a fine time generating them, with dice or spinners or slips of paper drawn from a sack).

The Monte Carlo method per se was invented and given its name at the Los Alamos Laboratory in the years after World War II. It's no surprise that the idea emerged there. They had big problems (even more frightening than collateralized mortgage obligations); they had access to early digital computers (ENIAC, MANIAC); and they had a community of creative problem-solvers (Stanislaw Ulam, John von Neumann, Nicholas Metropolis, Marshall Rosenbluth). From the outset, the Los Alamos group relied on pseudorandom numbers. At the first conference on Monte Carlo methods, in 1949, von Neumann delivered his famous quip, "Anyone who considers arithmetic methods of producing random digits is, of course, in a state of sin." Then he proceeded to sin.

Quasi–Monte Carlo was not far behind. The first publication on the subject was a 1951 report by Robert D. Richtmyer, who was head of the theoretical division at Los Alamos. The paper is an impressive debut. It sets forth the motivation for quasirandom sampling, introduces much of the terminology, and explains the mathematics. But it was also presented as an account of a failed experiment; Richtmyer had wanted to show improved convergence time for quasi–Monte Carlo computations, but his results were negative. I am a fervent believer in reporting negative results, but I have to concede that in this case the report may have inhibited further investigation.

In 1968 S. K. Zaremba, then at the University of Wisconsin in Madison, wrote a strident defense of quasirandom sampling (and a diatribe against pseudorandom numbers). As far as I can tell, he won few converts.

Work on the underlying mathematics of low-discrepancy sequences has gone on steadily through the decades (most notably, perhaps, by Klaus Friedrich Roth, I. M. Sobol, Harald Niederreiter, and Ian H. Sloan). Now there is renewed interest in applications, and not just among the Wall Street quants. It's catching on in physics and other sciences as well. Ray-tracing in computer graphics is another promising area.

The shifting fortunes of pseudo- and quasirandomness might be viewed in the context of larger trends. In the nineteenth century, randomness of any kind was an unwelcome intruder, reluctantly tolerated only where necessary (thermodynamics, Darwinian evolution). The twentieth century, in contrast, fell madly in love with all things random. Monte Carlo models were a part of this movement; the quantum theory was a bigger part, with its insistence on divine dice games. Paul Erdős introduced random choice into the methodology of mathematical proof, which is perhaps the unlikeliest place to look for it. In computing, randomized algorithms became a major topic of research. The concept leaked into the arts, too, with aleatoric music and the spatter paintings of Jackson Pollock. Then there was chaos theory. A 1967 essay by Alfred Bork called randomness "a cardinal feature of our century."

By now, though, the pendulum may be swinging the other way, most of all in computer science. Randomized algorithms are still of great practical importance, but the intellectual excitement is on the side of *derandomization*, showing that the same task can be done by a deterministic program. An open question is whether *every* algorithm can be derandomized. Deep thinkers believe the answer will turn out to be yes. If they're right, randomness confers no essential advantage in computing, although it may still be a convenience.

Quasirandomness seems to be steering us in the same direction, with a preference for taking our drams of randomness in the smallest possible doses. What's ahead may be a kind of homeopathic doctrine, where the greater the dilution of the active agent, the greater its potency. That's non-sense in medicine, but perhaps it works in mathematics and computation.

12

Pencil, Paper, and Pi

William Shanks was one of the finest computers of the Victorian era—when the term *computer* denoted not a machine but a person skilled in arithmetic. His specialty was mathematical constants, and his most ambitious project was a record-setting computation of the numerical value of π. Starting in 1850 and returning to the task at intervals over more than 20 years, he eventually published a value of π that began with the familiar digits 3.14159 and went on for 707 decimal places.

Seen from a twenty-first-century perspective, Shanks is a poignant figure. All his patient toil has been reduced to triviality. Anyone with a laptop can compute hundreds of digits of π in microseconds. Moreover, the laptop will give the correct digits. Shanks made a series of mistakes beginning around decimal place 530 that spoiled the rest of his work (see figure 12.1).

I have long been curious about Shanks and his 707 digits. Who was this prodigious human computer? What led him to undertake his quixotic adventures in arithmetic? How did he deal with the logistical challenges of the π computation: the teetering columns of figures, the grueling bouts of multiplication and division? And what went wrong in the late stages of the work?

One way to answer these questions would be to buy a ream of paper, sharpen a dozen pencils, and try to retrace Shanks's steps. I haven't the stamina for that, or even the life expectancy. But by adapting some pencil-driven algorithms to run on silicon computers, I have gotten a glimpse of what the process might have been like for Shanks. I think I also know where a couple of his errors crept in, but there are more that remain unexplained.

Scanty Intervals of Leisure

Biographical details about William Shanks are hard to come by. It's known that he was born in 1812, married in 1846, and died in 1882. He came from Corsenside, a village in the northeast of England, near the Scottish border. After his marriage he lived in Houghton-le-Spring, another small northeastern town, where he ran a boarding school.

Some sources identify Shanks as a student of William Rutherford, a mathematician who taught at the Royal Military Academy and also dabbled in π calculations. It's true that Shanks studied with Rutherford, but this was not the relationship of a graduate student with a thesis adviser. When Shanks published a small book on π in 1853, he dedicated it to Rutherford, "from whom I received my earliest lessons in numbers." It turns out that Rutherford taught at a school not far from Corsenside in the 1820s. Shanks was then a boy of 10 or 12, and he must have been one of Rutherford's pupils.

I have not been able to learn anything about Shanks's further education; there is no mention of a university degree. Rutherford remained a mentor and became a collaborator. The two men cross-checked their calculations of π and published some of the results jointly.

On balance, it seems that Shanks was an amateur and a marginal figure in the mathematical community, but not a crank. He published 15 papers in the *Proceedings of the Royal Society*. Although he was never a member, he apparently had no trouble persuading Fellows to submit manuscripts on his behalf. These sponsors, some of whom were also listed as subscribers to his 1853 book, included prominent figures in British science and mathematics: George Stokes, George B. Airy, William Whewell, Augustus De Morgan.

Pencil-and-paper computation was a prized skill in the nineteenth century. Compiling lists of prime numbers was respectable work for serious

Figure 12.1 Digits in the decimal expansion of π are plotted as dots, with digit values increasing from left to right and digit positions increasing from top to bottom. The *left panel* shows 707 digits, including both the correct values and those computed by William Shanks between 1850 and 1873. Where the two values agree, the dots are gray; where they disagree, the correct values are black and Shanks's are white. The *right panel* shows digits 500 through 707 in greater detail, with horizontal lines connecting the discrepant values. Errors begin in the 528th decimal place.

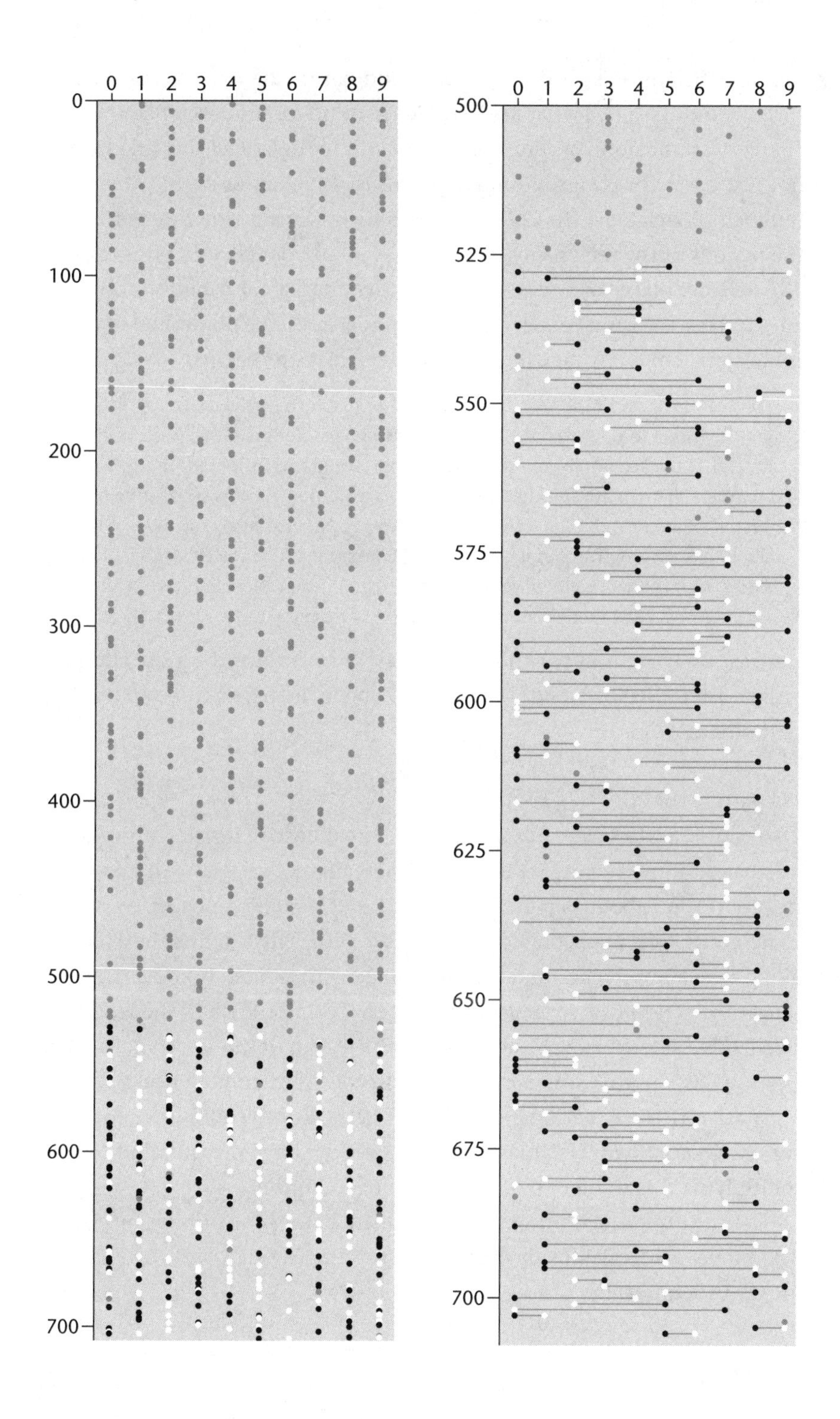
0
1
2
3
4
5
6
7
8
9
0
100
200
300
400
500
600
700
500
525
550
575
600
625
650
675
700

mathematicians. (Gauss did it.) The making of mathematical tables was a thriving industry. (It's what Babbage had in mind for his difference engine.) Shanks was not alone in computing constants to high precision, and he was not just a π man. He also found high-precision values of e and γ (both are numbers associated with Euler), logarithms of several small integers, and the periods of the reciprocals of all primes up to 60,000. Still, grinding out 707 decimal places of π was more of a stunt than a contribution to mathematical research. Shanks seems to have understood the borderline status of his project. The book he wrote about his calculations begins,

> Towards the close of the year 1850, the Author first formed the design of rectifying the Circle to upwards of 300 places of decimals. He was fully aware, at that time, that the accomplishment of his purpose would add little or nothing to his fame as a Mathematician, though it might as a Computer; nor would it be productive of anything in the shape of pecuniary recompense at all adequate to the labour of such lengthy computations. He was anxious to fill up scanty intervals of leisure with the achievement of something original, and which, at the same time, should not subject him either to great tension of thought, or to consult books.

He was surely right about the limited payoff in fame and funds. The subscriber list attests to sales of 36 copies. I hope he managed to avoid tension of thought.

The Recipe for Pi

There are countless ways of computing π, but almost all nineteenth-century calculators chose arctangent formulas. These methods begin with a geometric observation about a circle with radius 1 and circumference 2π. As shown in figure 12.2, an angle drawn at the center of the circle defines both an arc along the circumference and a right triangle with sides a, b, and c. The arctangent function (the inverse of the tangent function) relates the length of side b (the "side opposite" the angle) to the length of the arc. In particular, when b has length 1, the arc is one-eighth of the circumference, which is equal to $\pi/4$. The equation $\arctan 1 = \pi/4$ is the key to computing π. If you can assign a numerical value to arctan 1, you get an approximation to $\pi/4$; multiplying this number by 4 yields a value for π itself.

The next question is how to compute an arctangent to high precision. The pioneers of calculus devised an infinite series that gives the value of $\arctan x$ for any value of x between -1 and $+1$:

$$\arctan x = \frac{x^1}{1} - \frac{x^3}{3} + \frac{x^5}{5} - \frac{x^7}{7} + \cdots.$$

For the case of $x = 1$, the series assumes a particularly simple form:

$$\arctan 1 = \frac{1}{1} - \frac{1}{3} + \frac{1}{5} - \frac{1}{7} + \cdots.$$

Hence, to calculate $\pi/4$, one can just add up the terms of this series—the reciprocals of successive odd numbers, with alternating plus and minus signs—until the sum attains the desired accuracy.

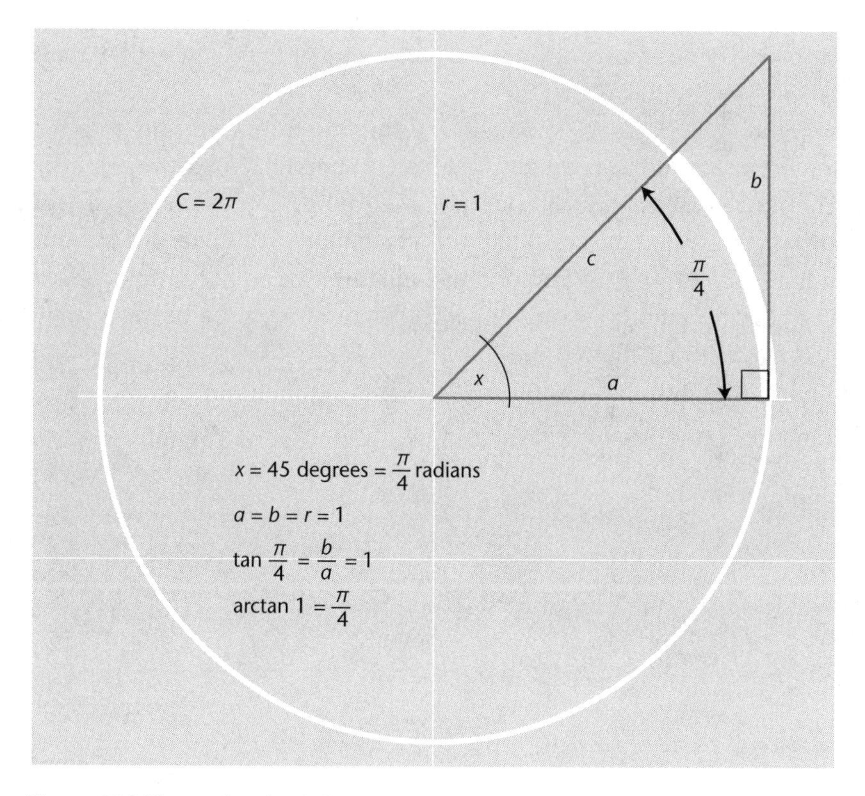

Figure 12.2 The pie slice that helps determine the value of π is one-eighth of a circle, with an angle of 45 degrees, or $\pi/4$ radians. The tangent of this angle, defined as the ratio b/a in the gray triangle, is equal to 1. Hence, computing the arctangent of 1 yields a numerical value for $\pi/4$.

Lamentably, this plan won't work. At $x = 1$ the arctan series converges at an agonizingly slow pace. To get n digits of π, you need to sum roughly 10^n terms of the series. Shanks would have had to evaluate more than 10^{700} terms, which was beyond the means of even the most intrepid Victorian computer.

All is not lost. For values of x closer to zero, the arctan series converges more quickly. The trick, then, is to combine multiple arctan calculations that sum up to the same value as arctan 1. Shanks worked with a formula discovered in 1706 by the English mathematician John Machin:

$$\frac{\pi}{4} = 4 \arctan\frac{1}{5} - \arctan\frac{1}{239}.$$

He had to evaluate two arctan series rather than just one, but both of these series converge much faster.

Figure 12.3 traces the evaluation of the first three terms of the series for arctan 1/5 and arctan 1/239, retaining five decimal places of precision. The error in the computed value of π is 0.00007. No extraordinary skill in arithmetic is needed to carry out this computation by hand. But imagine scaling it up to several hundred terms and several hundred decimal places. The basic operations remain the same, but keeping all the figures straight becomes a clerical nightmare.

$\arctan x$	=	$\frac{x^1}{1}$	−	$\frac{x^3}{3}$	+	$\frac{x^5}{5}$		
$\arctan \frac{1}{5}$	=	0.20000	−	0.00267	+	0.00006	=	0.19739
							×	4
								0.78956
$\arctan \frac{1}{239}$	=	0.00418	−	0.00000	+	0.00000	=	− 0.00418
								0.78538
							×	4
				$\pi = 4\left(4 \arctan \frac{1}{5} - \arctan \frac{1}{239}\right)$			≈	3.14152

Figure 12.3 A crude computation of π proceeds by summing the first three terms in an infinite series for arctan 1/5 and arctan 1/239. In the series for arctan 1/5, the first term is simply (1/5)/1, and the second term is $(1/5)^3/3$. Each term is evaluated to five decimal places. Plugging these values into John Machin's formula (*last row*) yields four correct decimal places of π.

```
0.2000000000 0000000000 0000000000 0000000000 0000000000 0000000000 0000000000 0000000000 0000000000 0000000000
0.0000640000 0000000000 0000000000 0000000000 0000000000 0000000000 0000000000 0000000000 0000000000 0000000000
0.0000000568 8888888888 8888888888 8888888888 8888888888 8888888888 8888888888 8888888888 8888888888 8888888888
0.0000000000 6301538461 5384615384 6153846153 8461538461 5384615384 6153846153 8461538461 5384615384 6153846153
0.0000000000 0007710117 6470588235 2941176470 5882352941 1764705882 3529411764 7058823529 4117647058 8235294117
0.0000000000 0000009986 4380952380 9523809523 8095238095 2380952380 9523809523 8095238095 2380952380 9523809523
0.0000000000 0000000013 4217728000 0000000000 0000000000 0000000000 0000000000 0000000000 0000000000 0000000000
0.0000000000 0000000000 0185127900 6896551724 1379310344 8275862068 9655172413 7931034482 7586206896 5517241379
0.0000000000 0000000000 0000260301 0482424242 4242424242 4242424242 4242424242 4242424242 4242424242 4242424242
0.0000000000 0000000000 0000000371 4566310054 0540540540 5405405405 4054054054 0540540540 5405405405 4054054054
0.0000000000 0000000000 0000000000 5363471355 0048780487 8048780487 8048780487 8048780487 8048780487 8048780487
0.0000000000 0000000000 0000000000 0007818749 3530737777 7777777777 7777777777 7777777777 7777777777 7777777777
0.0000000000 0000000000 0000000000 0000011488 7745596186 1224489795 9183673469 3877551020 4081632653 0612244897
0.0000000000 0000000000 0000000000 0000000016 9947155749 8300377358 4905660377 3584905660 3773584905 6603773584
0.0000000000 0000000000 0000000000 0000000000 0252833663 2909752140 3508771929 8245614035 0877192982 4561403508
0.0000000000 0000000000 0000000000 0000000000 0000378007 0506907695 0032786885 2459016393 4426229508 1967213114
0.0000000000 0000000000 0000000000 0000000000 0000000567 5921253449 0928049230 7692307692 3076923076 9230769230
0.0000000000 0000000000 0000000000 0000000000 0000000000 8555011744 3290674161 1594202898 5507246376 8115942028
0.0000000000 0000000000 0000000000 0000000000 0000000000 0012937990 3640264252 4300273972 6027397260 2739726027
0.0000000000 0000000000 0000000000 0000000000 0000000000 0000019625 4191495881 3595302233 7662337662 3376623376
0.0000000000 0000000000 0000000000 0000000000 0000000000 0000000029 8500202373 9825122731 2987654320 9876543209
0.0000000000 0000000000 0000000000 0000000000 0000000000 0000000000 0455125014 4431545128 3056037647 0588235294
0.0000000000 0000000000 0000000000 0000000000 0000000000 0000000000 0000695471 9321827979 0724669900 2247191011
0.0000000000 0000000000 0000000000 0000000000 0000000000 0000000000 0000001064 8946574497 8948378419 2880860215
0.0000000000 0000000000 0000000000 0000000000 0000000000 0000000000 0000000001 6335703611 1885232151 6370110020
```

Figure 12.4 Merely writing out all the terms of the arctan series and summing them up must have been a challenge for Shanks. Shown here are the first 25 positive terms of the series for arctan 1/5, each computed to 100-digit precision, and printed in minuscule numerals. Shanks calculated 253 positive terms, each to 709 digits. Written at normal size, they would have filled a sheet of paper 2 meters wide and 1 meter high.

In computing arctan 1/5, Shanks evaluated 506 terms, each carried to 709 decimal places. Most likely he performed separate summations of the positive and negative terms. If he had tried to write down such an addition problem all in one piece—253 rows of 709-digit numbers, or almost 180,000 digits in all—it would have filled a sheet of paper 2 meters wide by 1 meter high (see figure 12.4). Was paper made in such large sheets in the 1870s? I have found one source that lists a size called Emperor, 6 feet by 4 feet. Of course, Shanks could have pasted together smaller sheets, or he could have broken the task down into smaller subproblems that could be handled without special-order stationery. That would have made the arithmetic less awkward physically, but it would have entailed other costs: copying intermediate results from one sheet to another, transferring carry digits, the risk of misaligning columns or rows.

Erwin Engert, a Shanks enthusiast, has tested the travails of pencil-and-paper calculation by doing 20-digit, 40-digit, and 100-digit evaluations of Machin's arctan formula. During this work the challenge of keeping digits aligned became severe enough that Engert printed ruled forms for the larger computations. Shanks may well have done the same, although we have no direct evidence.

Pencil-Friendly Algorithms

In silico, summing n terms of the series for arctan x takes just a few lines of code:

```
function arctan(x, n)
   sum = 0
   for k from 0 to n-1
      sign = (-1)^k
      m = 2 * k + 1
      term = sign * (x^m)/m
      sum = sum + term
   return sum
```

For each integer k from 0 to $n-1$, the program generates an odd integer m and the corresponding term of the arctan series, x^m/m. The expression $(-1)^k$ sets the sign of the term—plus for even k, minus for odd. When the loop completes, the function returns the accumulated sum of the n terms. The only hidden subtlety here is that the numeric variables must be able to accommodate numbers of arbitrary size and precision.

No one doing arithmetic with a pencil would adopt this algorithm. After every pass through the loop, the program throws away all its work except the variables k and *sum*, then starts from scratch to build the next term of the series. A manual worker would surely save the value of x^m as a starting point for calculating the next power, x^m+2. And exponentiating –1 is not how a human computer would keep track of alternating signs.

It's not hard to transform the program into a more pencil-friendly procedure, avoiding needless recomputation and saving intermediate results for future use. The computer could also be programmed to use digit by digit algorithms—the ones we all learned in elementary school, and forgot soon after—for multiplication and long division. But these steps still fail to capture some important practices of a shrewd human reckoner.

Most of the terms in the series for arctan 1/5 are repeating decimals with a short period. For example, the term $(1/5)^9/9$ works out to 0.000000056888. . . . A naive computer program would go on dividing digit after digit out to the limit of precision, but Shanks surely just filled in a string of 8s.

Peculiarities of base 10 also offer certain shortcuts. For generating the sequence of odd powers of 1/5, the basic step is dividing by 25. Engert suggests dividing by 100 (a shift of the decimal point by two places) and multiplying by 4. Another option is to calculate $(1/5)^m$ as $2^m/10^m$ (where, again, division by a power of 10 is just a shift of the decimal point). I mention this latter possibility because Shanks's book on the π computation includes a table of the powers of 2 up to 2^{721}. Did he use those numbers to compute his powers of 1/5, or were they just for checking values computed in some other way?

Shanks doesn't reveal much about his methods, and I remain unsure about several aspects of his strategy. For example, a term in the series for arctan 1/5 can be written either as $(1/5)^m/m$ or as $1/(m5^m)$. Mathematically these expressions are identical, but they imply different computations. In the first case you multiply and divide long decimal fractions; in the second you build a large integer and then take its reciprocal. Which way did Shanks do it? He doesn't say. If I were to attempt to replicate his work, I might stick with decimal fractions for arctan 1/5 because of the many short-period repetitions, but I might choose the reciprocal method for arctan 1/239 because taking a reciprocal is a little easier than other forms of division.

Where He Went Wrong

As Tolstoy might have said, all correct computations are alike, but every erroneous computation errs in its own way. In that spirit, the incorrect digits in Shanks's result are much more informative than the correct ones. They might reveal where and how his calculation went off the rails.

Shanks published his value of π in three stages. A January 1853 article (under Rutherford's byline) includes 530 decimal places; 440 of those figures were confirmed by Rutherford, and all of them are correct apart from a few typographical errors and a discrepancy in the last two digits that could be attributed to round-off. In the spring of 1853 Shanks extended his calculation from 530 to 607 decimal places, publishing these results in a privately printed book, *Contributions to Mathematics, Comprising Chiefly the Rectification of the Circle to 607 Places of Decimals*. This is where the errors creep in. There's evidence that he made at least four mistakes.

After bringing out his book, Shanks put π aside for 20 years. When he took up the task again in 1873, he extended the two arctan series to 709 decimal places and π to 707. Because these computations were built atop the flawed earlier work, they were doomed from the start. Yet none of the errors were noticed until 75 years later, when D. F. Ferguson, working with a mechanical desk calculator, extended a new calculation of π beyond 700 digits.

Trying to discover where Shanks went wrong is an interesting exercise in forensic mathematics. Usually, one strives to find the correct answer to a problem; here the aim is to get the wrong answer—but the *right* wrong answer. We want to take a correct value and find some way of modifying it that will yield the specific erroneous output reported by Shanks. It's like searching for a suspicious transaction when your checkbook disagrees with the bank statement, except that we have no access to the individual checkbook entries, only the final balance.

When Shanks published his initial 530-digit computation, he included all the individual terms of the two arctan series. For the extensions to 607 and 707 digits, however, he published only his final values of arctan 1/5, arctan 1/239, and π. This is unfortunate; if we had his working papers, locating the source of his errors would be much easier. As things stand, the only evidence in hand consists of three huge numbers, and the knowledge that they differ in their low-order digits from the correct values.

In tracing where Shanks went astray, the obvious way to start is to take the difference between the true value of π and his result:

$$
\begin{array}{ll}
T_\pi & 3.141 \ldots 08602139494639522473 7 \\
S_\pi & 3.141 \ldots 086021395\underline{016092448077} \\
\Delta_\pi & -0.000 \ldots 000000000069697223340
\end{array}
$$

Here T_π is the true π sequence, showing the first few digits and then digits 520 through 540. S_π lists the corresponding digits from Shanks's final publication on π, with the underline indicating the start of the region where the computation goes awry, at decimal place 528. Finally, Δ_π records the difference between these two numbers.

This discrepancy must reflect a similar error in one of the two arctan series. Shanks's value for arctan 1/239 shows no anomalies in the neighborhood of digit 528, but the arctan 1/5 series is flawed in exactly the way needed to explain the error at the 528th decimal place in Shanks's value for π. Again, we compare digits 520 through 540:

$$\begin{array}{ll} T_{1/5} & 0.197\ \ldots\ 756051837757422087783 \\ S_{1/5} & 0.197\ \ldots\ 7560518377\underline{61778164242} \\ \Delta_{1/5} & -0.000\ \ldots\ 000000000004356076458 \end{array}$$

In this case the aberrant digits begin not at position 528 but at 530, which means the magnitude of the error is smaller. This is exactly what is to be expected. In Machin's formula, the value of arctan 1/5 is multiplied by 4 to get a value for $\pi/4$, then that result is multiplied by 4 again to get the value of π itself. And if you multiply the error term $\Delta_{1/5}$ by 16, you get the corresponding error in Shanks's value of π, namely, Δ_{π}.

From this analysis we can infer that Shanks almost surely made a mistake somewhere around digit 530 in his computation of arctan 1/5. It's worth noting that this location lies right at the boundary between two stages in his work. Early in 1853 he had completed and published 530 digits of π, then a few weeks later he took up the task anew, extending his results to 607 digits. Perhaps that transition was a particularly perilous moment in the project, as he grafted new results onto the old paperwork.

Since the arctan 1/5 value is wrong, at least one of the 506 terms contributing to that value must also be flawed, or else Shanks added them incorrectly. In principle, there could be multiple mistakes in multiple terms, but in that case the quest to reconstruct the computation is probably hopeless. So, on the principle of looking for lost keys under the streetlight, it seems best to focus on common kinds of errors. Maybe Shanks omitted a term entirely when he was adding up that tower of 506 numbers. Or perhaps he truncated a term, neglecting to include the new digits from position 530 on. Or in piecing together old and new work he could have omitted a few digits somewhere, or inserted some extraneous ones.

As it happens, errors like these are not only easy to make but also easy to detect. Accordingly, I wrote a series of programs to search for them in the 506 terms. My aim was to "uncorrect" the true value of each term by subtracting the known error $\Delta_{1/5}$, then look for the signature of some simple transformation that would account for the difference. For example, if a term was truncated at decimal place 530, the "uncorrected" version would have a string of zeros starting at that position.

The search was soon rewarded. In term 497 of the arctan series (that is, the term computing $(1/5)^{497}/497$) I discovered a highly suspicious pattern:

T_{497} 0.000 ... 907444668008$\underline{0}$482897384
$\Delta_{1/5}$ −0.000 ... 0000000000004356076458
U_{497} 0.197 ... 9074446680084838973843

T_{497} is the correct value of the term; subtracting $\Delta_{1/5}$ produces the "uncorrected" value U_{497}. What's intriguing in these numbers is that deleting the underlined zero brings the correct and the uncorrected sequences into alignment, or nearly so. (A single-digit substitution is also needed.) A diagram makes it clearer what's going on:

delete
9 0 7 4 4 4 6 6 8 0 0 8 $\underline{0}$ 4 8 2 8 9 7 3 8 4
↓ ↓ ↓ ↓ ↓ ↓ ↓ ↓ ↓ ↓ ↓ ↓ / / / / / / / / / shift
9 0 7 4 4 4 6 6 8 0 0 8 4 8 2 8 9 7 3 8 4 3
↓substitute
9 0 7 4 4 4 6 6 8 0 0 8 4 8 3 8 9 7 3 8 4 3

From this observation comes a causal hypothesis: Shanks inadvertently omitted the underlined 0, and all the following digits shifted left by one place. In a separate error Shanks apparently substituted a 3 for a 2. If you inject the one-digit shift error into the arctan calculation and also make the substitution, the output exactly matches the Shanks value in the region following decimal place 530.

Without further documentary evidence, it's not possible to prove that this spot marks the site of Shanks's first error, but it's certainly plausible. When Shanks extended this term from 530 digits to 609, he didn't need to do any actual arithmetic. Term 497 is a repeating decimal with a period of 210 digits, so he merely needed to copy a segment from earlier in the sequence. It seems likely that he missed that 0 digit while copying.

I was not the first to discover this error; Engert identified it before I did.

Further Wrong Turns

If the delete-shift error in term 497 were Shanks's only misstep, then inserting that change into the computation would reproduce all 707 digits of the Shanks value for π. But matters are not so simple. The uncorrection produces about 40 of the right wrong digits, but then at decimal place 569 the two sequences part ways again. Evidently there's another mistake. Tracking down the cause of this second error was an adventure that took me into some unexpected corners of the mathematical universe. Along the way I took a few wrong turns of my own.

The first step, again, is to find the difference between the true value of arctan 1/5 and the Shanks value, but now the true value is no longer true; it's the "uncorrected" number $U_{1/5}$ that incorporates an adjustment for the earlier error at decimal place 530. Subtracting the Shanks result $S_{1/5}$ from $U_{1/5}$ gives a new error term $\Delta_{1/5}$. The underlined region of $S_{1/5}$ begins at digit 569:

$U_{1/5}$	0.197 ... 0282776862915647787102
$S_{1/5}$	0.197 ... 0282776861191509856067
$\Delta_{1/5}$	0.000 ... 0000000001724137931034

The difference $\Delta_{1/5}$ can now be applied to all 506 terms of the arctan sum to see if anything interesting turns up.

A note in the 1946 paper by D. F. Ferguson that first reported Shanks's errors led me to examine term 145. Sure enough, I immediately spotted a pattern suggesting another delete-shift transformation:

U_{145}	0.000 ... 4827586206896551724137
$\Delta_{1/5}$	0.000 ... 0000000001724137931034
UU_{145}	0.000 ... 4827586205172413793103

UU_{145} is the doubly adjusted term, incorporating uncorrections for the errors at both digit 530 and digit 569. It can be generated by omitting the five underlined digits in U_{145} and shifting the rest five positions to the left.

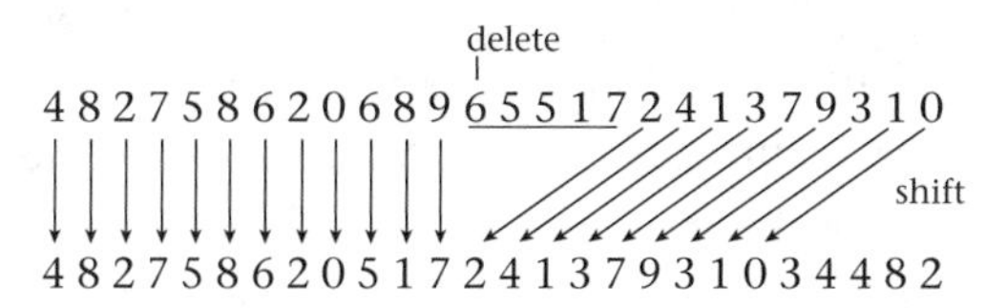

Once again, it's easy to imagine a careless error, perhaps while transcribing intermediate results.

Does that solve the mystery of error no. 2? Maybe, but something about those numbers—those particular long strings of digits—bothered me. Here's a slightly larger sample of U_{145}, $\Delta_{1/5}$, and UU_{145} with different underlining:

413793103448275862068965517241379310344827586206896551724137
000000000000000000017241379310344827586206896551724190124797
413793103448275862051724137931034482758620689655172361599340

The very same 28-digit motif appears in all three numbers, shifted left or right by a few places. The presence of this sequence in term 145 is easy to explain. These 28 digits form the repeating unit, or *repetend*, in the decimal representation of $(1/5)^{145}/145$. If you were to compute still more digits of the term, you would find that the sequence repeats endlessly. But seeing it turn up in the other two numbers seems a bit surprising.

Try an experiment. Take the sequence 1724137931034482758620689655 and create a cyclic permutation. For any k between 1 and 27, lop off k digits from the front of the sequence and paste them onto the back. Now subtract the permuted sequence from the original one. Here's an example with $k = 7$:

```
 1724137931034482758620689655
−9310344827586206896551724137
−7586206896551724137931034482
```

Look closely and you'll see that the digits in the result form another cyclic permutation of the same sequence. That can't be a coincidence!

And indeed it's not. The sequence 1724137931034482758620689655 is a cyclic permutation of 0344827586206896551724137931, which is the repetend in the decimal expansion of 1/29. Note that 145, the denominator in the term, is equal to 5×29. The secret behind all this hocus-pocus is that 29 is a full-repetend prime in base 10, a prime p whose reciprocal $1/p$ has a repetend with $p-1$ digits, the most possible. For all such primes the repetend is a cyclic number: multiplying by any integer in 1, 2, 3 … $p-1$ yields a cyclic permutation of the same digits. And, as observed, subtracting any two cyclic permutations yields another cyclic permutation.

The extraordinary properties of this cyclic number inspired me to have a look at term 29 in the arctan series. This is what I found for the "uncorrected" true value, the Δ, and the predicted Shanks value for term 29:

4137931034482758620689655172413793103448275862068965517241
0000000000000000000000000172413793103448275862068965517241
4137931034482758620689654999999999999999999999999999999999

The terminal sequence 4999... could just as well be 5000...; the two forms are equivalent. This points to another possible explanation of the error at decimal place 569: Shanks may have truncated term 29 at this position, neglecting to include all the remaining digits in the sum.

Two theories are not always better than one. Did Shanks skip five digits in term 145, or did he drop the trailing digits of term 29? Since those two acts lead to the same outcome, it's hard to resolve the question. And Ferguson's note from 1946 adds further confusion. He speaks of truncating term 145 (which does *not* yield the observed result), and says nothing about term 29.

In any case, this is not the last of Shanks's blunders. In the computation of arctan 1/5 there is at least one more muddle at digit 602. The last eight digits of the 609-place value published in 1853 differ from the corresponding digits listed in 1873, and both differ from the true value. I have not found a simple error that yields either of the Shanks versions. There's also at least one mistake in the arctan 1/239 calculation, which I have not tried to track down.

It's curious that Shanks produced almost 530 flawless digits of π, then made at least four mistakes in the next 80 digits. All four errors date from March or April of 1853, and they seem to be clerical rather than mathematical. I can only speculate on the cause of this sudden spate of carelessness. Perhaps Shanks was hurrying to get his book into the hands of the subscribers. Or maybe, at age 41, he was experiencing the early symptoms of presbyopia.

Stories about Shanks tend to focus on the mistakes. We look back with pity and horror on all those pages of meticulous arithmetic rendered worthless by a slip of the pencil. But I would argue that even with the errors, Shanks's computation of π was an impressive endeavor. His 527 correct digits were not bettered for almost a century. Augustus De Morgan, one of the leading mathematicians of the era, had his doubts about Shanks's work, but he also spoke admiringly of "the power to calculate, and . . . the courage to face the labour."

13

Foolproof

I was a teenage angle trisector. In my first full-time job, fresh out of high school, I trisected angles all day long for $1.75 an hour. My employer was a maker of voltmeters, ammeters, and other electrical instruments. This was back in the analog age, when a meter had a slender pointer swinging in an arc across a scale (see figure 13.1). My job was drawing the scale. A technician would calibrate the meter, recording the pointer's angular deflection at a few key intervals—say, 3, 6, 9, 12, and 15 volts. When I drew the scale, using ruler and compass and a fine pen, I would fill in the intermediate divisions by interpolation. That's where the trisection of angles came in. I was also called upon to perform quintisections and various other impossible feats.

I joked about this with my coworker and supervisor, Dmytro, who had been drawing meter scales for some years. We should get extra pay, I said, for solving one of the famous unsolvable problems of antiquity. But Dmytro was a skeptic, and he challenged me to prove that trisection is impossible. This was beyond my ability. I did my best to present an outline of a proof (after rereading a Martin Gardner column on the topic), but my grasp of the mathematics was tenuous, my argument was incoherent, and my audience remained unconvinced.

On the other hand, Dmytro himself quickly produced visible evidence that the specific method of trisection we employed—drawing a chord across the angle and dividing it into three equal segments—gave incorrect results when applied to large angles (see figure 13.2). After that, we made sure all the angles we trisected were small ones. And we agreed that the whole matter was something we needn't discuss with the boss. Our circumspect silence was a little like the Pythagorean conspiracy to conceal the irrationality of $\sqrt{2}$.

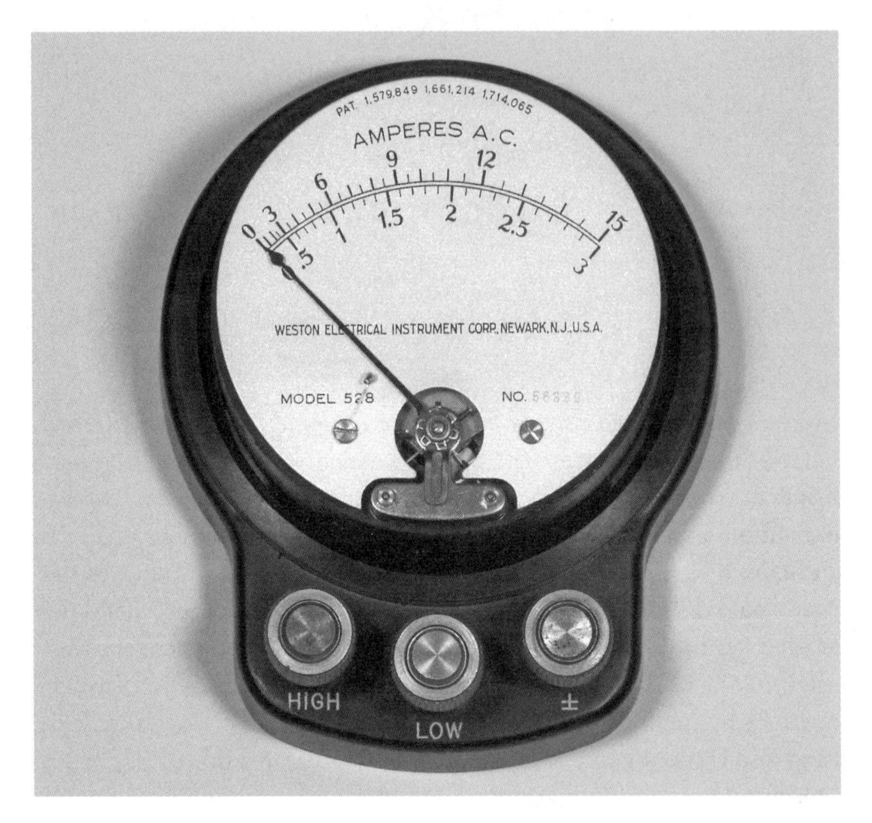

Figure 13.1 Drawing scales for instruments such as ammeters sometimes required trisection of angles to interpolate between major divisions. The meter scale shown here dates from the era when I was performing trisections, but it is not my work, and it shows evidence of an algorithm more sophisticated than simple interpolation.

Looking back on this episode, I am left with vague misgivings about the place of proof in mathematical discourse and in everyday life. Admittedly, my failure to persuade Dmytro was entirely a fault of the prover, not of the proof. Still, if proof is a magic wand that works only in the hands of wizards, what is its utility to the rest of us?

Reading Euclid Backward

Here is how proof is supposed to work, as illustrated by an anecdote in John Aubrey's *Brief Lives* about the seventeenth-century philosopher Thomas Hobbes:

> He was 40 yeares old before he looked on geometry; which happened accidentally. Being in a gentleman's library in . . . , Euclid's Elements lay open, and 'twas the 47 *El. libri* I. He read the proposition. "By G—," sayd he (he would now and then sweare, by way of emphasis), "this is impossible!" So he reads the demonstration of it, which referred him back to such a proposition; which proposition he read. That referred him back to another, which he also read. *Et sic deinceps,* that at last he was demonstratively convinced of that trueth. This made him in love with geometry.

What's most remarkable about this tale—whether or not there's any trueth in it—is the way Hobbes is persuaded against his own will. He starts out incredulous, but he can't resist the force of deductive logic. From proposition 47 (which happens to be the Pythagorean theorem), he is swept backward through the book, from conclusions to their premises and eventually to axioms. Though he searches for a flaw, each step of the argument compels assent. This is the power of pure reason.

For many of us, the first exposure to mathematical proof—typically in a geometry class—is rather different from Hobbes's middle-age epiphany. A nearer model comes from another well-worn story, found in Plato's dialogue *Meno.* Socrates, drawing figures in the sand, undertakes to coach an

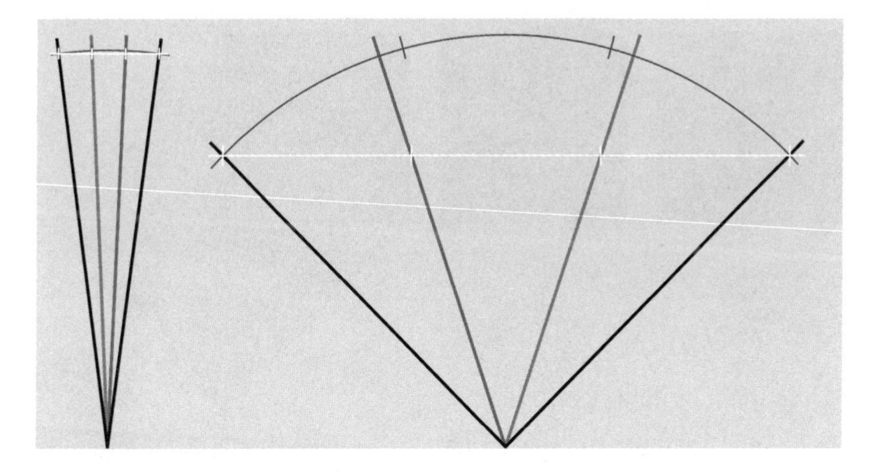

Figure 13.2 One crude trisection method draws a chord across the angle and divides it into three equal segments. The result is an approximation, good for narrow angles but visibly inaccurate for wider ones. The trisected chords are shown in white, and correct trisections are indicated on the gray arcs.

untutored slave boy, helping him to prove a special case of the Pythagorean theorem (see figure 13.3). I paraphrase very loosely:

Socrates: Here is a square with sides of length 2 and area equal to 4. If we double the area, to 8 units, what will the length of a side be?

Boy: Umm, 4?

Socrates: Does $4 \times 4 = 8$?

Boy: Okay, maybe it's 3.

Socrates: Does $3 \times 3 = 8$?

Boy: I give up.

Socrates: Observe this line from corner to corner, which the erudite among us call a *diagonal*. If we erect a new square on the diagonal, note that one-half of the original square makes up one-fourth of the new square, and so the total area of the new square must be double that of the original square. Therefore the length of the diagonal is the length we were seeking, is it not?

Boy: Whatever.

At this point I trust we are all rooting for the kid. I would like to be able to report that the dialogue continues with the boy taking the initiative, saying something like, "Okay, dude, so what's the length of your erudite diagonal?

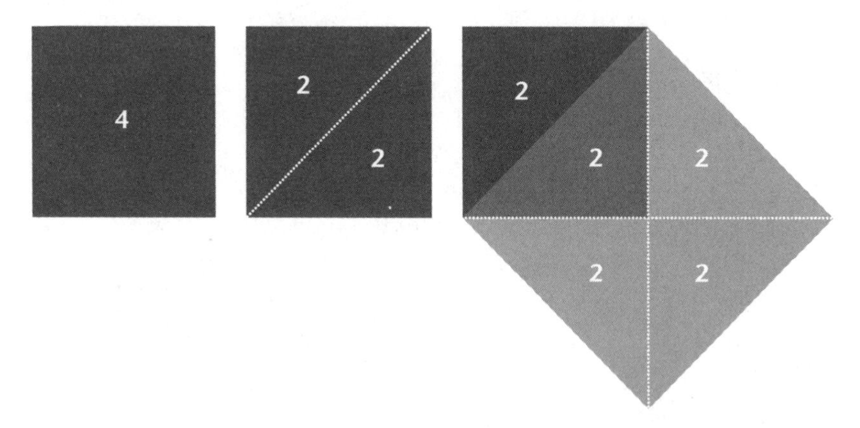

Figure 13.3 Doubling of a square is a proof taught to a slave boy in Plato's *Meno*. A simplified version of the proof begins with a square of side 2 and area 4, bisects the square along the diagonal, and constructs a new square whose side is the diagonal of the original square. The isosceles triangle forming half the area of the smaller square is one-fourth the area of the larger one, and so the area has doubled.

It's not 4 and it's not 3, so what is it, exactly?" Alas, Plato reports no such challenge from the slave boy.

The problem with the *Meno* proof is the opposite of the one I faced when I was an untutored wage slave. Whereas I was too inept and intellectually ill-equipped to craft a proof that would persuade my colleague (or even myself), Socrates is a figure of such potent authority that the poor kid would surely go along with anything the master said. He would put up no resistance even if Socrates were proving that 1 = 2. It's hard to believe that the boy will go on to prove theorems of his own.

Sadly, Hobbes didn't get much more benefit from his own geometry lesson. He became a notorious mathematical crank, claiming to have solved all the most famous problems of classical geometry, including the trisection of the angle, the squaring of the circle, and the doubling of the cube. These claims were a little less foolish in the seventeeth century than they would be now, since the impossibility of the tasks had not yet been firmly established. Nevertheless, Hobbes's contemporaries had no trouble spotting the gaffes in his purported proofs.

Enormous Theorems, Unwieldy Proofs

In recent years proof has become a surprisingly contentious topic. One thread of discord began with the 1976 proof of the four-color-map theorem by Kenneth Appel, Wolfgang Haken, and John Koch of the University of Illinois at Urbana-Champaign. They showed that if you want to color a map so that no two adjacent countries share a color, four crayons are all you'll need. The proof relied on computer programs to check thousands of map configurations. This intrusion of the computer into pure mathematics was greeted with suspicion and even disgust. Haken and Appel reported a friend's comment: "God would never permit the best proof of such a beautiful theorem to be so ugly." Apart from such emotional and aesthetic reactions, there was the nagging question of verification. How could we ever be sure the computer didn't make a mistake?

Some of the same issues came up again with the proof of the Kepler conjecture by Thomas C. Hales of the University of Pittsburgh (with contributions by his student Samuel P. Ferguson). The Kepler conjecture says that the pyramid of oranges on a grocer's shelf is packed as densely as possible. Computer calculations played a major part in the proof. Although this reliance on technology did not evoke the same kind of revulsion expressed two decades before, worries about correctness had not gone away.

Hales pronounced his proof complete in 1998, submitting six papers for publication in *Annals of Mathematics*. The journal enlisted a dozen referees to examine the papers and their supporting computer programs, but in the end the reviewers were defeated by the task. They had found nothing wrong, but the computations were so vast and formless that exhaustive checking was impractical, and the referees felt they could not certify the entire proof to be error-free. This was a troubling impasse. Eventually the *Annals* published "the human part of the proof," excluding some of the computational work. Hales then set to work rewriting the proof in a formal logic notation that could be checked by a computer. It took more than ten years; he announced the successful completion of the project in 2014.

If some proofs are too long to comprehend, others are too terse and cryptic. In 2003 the Russian mathematician Grigory Perelman announced a proof of the Poincaré conjecture. This result says—here I paraphrase Christina Sormani of Lehman College of the City University of New York—that if a blob of alien goo can ooze its way out of any lasso you throw around it, then the blob must be nothing more than a deformed sphere, without holes or handles. Everyday experience testifies to this fact for two-dimensional surfaces, and the conjecture was proved some time ago for surfaces (or manifolds) of four or more dimensions. The hard case was the three-dimensional manifold, which Perelman solved.

Perelman's proof is not easy reading. Sormani explains its strategy as "heating the blob up, making it sing, stretching it like hot mozzarella and chopping it into a million pieces." I applaud the vividness of this description, and yet it has not helped me follow Perelman's logic. Given the difficulty of the work, others stepped in to explicate and elaborate, publishing versions of the proof considerably longer than the original. These were not popularizations aimed at the general public; they were meant to explain the mathematics to mathematicians. Controversy ensued. Were the explicators trying to claim a share of the glory? Did they *deserve* a share? In the end it was Perelman who was awarded a Fields Medal, the biggest prize in mathematics, and a $1 million Millennium Prize. (He refused both.)

Now the mathematics community is coping with another problematic proof. The *abc* conjecture, formulated in the 1980s, sounds like an innocent,

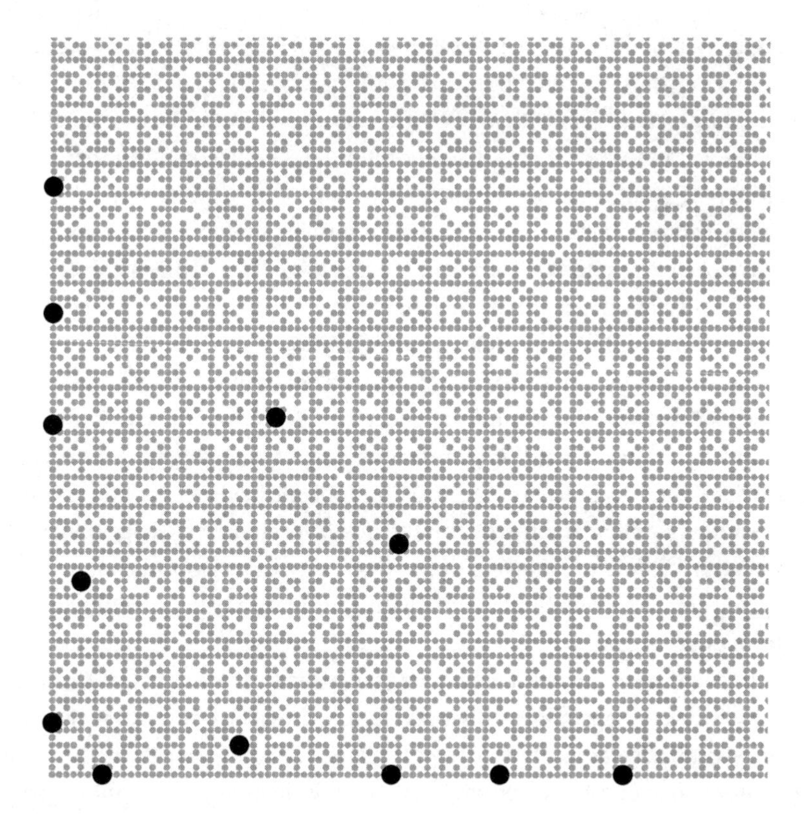

Figure 13.4 The *abc* conjecture is a simple statement about the additive and multiplicative properties of whole numbers. The small gray dots represent all pairs of numbers between 1 and 100 that have no prime factors in common; the large black dots form a subset of these numbers called "*abc* hits." The conjecture states that *abc* hits are rare, which one might surmise from the sparse sprinkling of black dots in the diagram. A proposed proof, however, is so impenetrable it is testing the ability of the mathematics community to digest it.

elementary problem in number theory that might be suitable for high school students (see figure 13.4). We begin with two integers, a and b, that have no prime factors in common. In other words, no number divides both a and b. Now calculate the sum $a+b=c$, then list the prime factors of all three numbers and cast out any duplicates; the product of the distinct factors is designated R. If $R < c$, we have an "*abc* hit." The conjecture says that *abc* hits are rare. The only subtlety in the problem statement is defining what rare means.

As an example of the *abc* process, take $a = 5$, $b = 27$, with sum $c = 32$. The prime factors of those three numbers are 2, 2, 2, 2, 2, 3, 3, 3, 5. After removing duplicates, we're left with a set of three primes: {2, 3, 5}. Their product R is equal to 30. Because $R < c$, this is an *abc* hit. But it's easy to confirm that $a = 5$, $b = 16$, $c = 21$ is *not* a hit.

Such a simply stated problem begs for a simple proof, but that is not what we have been given. In 2012 Shinichi Mochizuki of Kyoto University released four papers totaling more than 500 pages that purport to prove the *abc* conjecture. (With revisions and addenda, the page count is now over 1,000.) The proof is not only long but also formidably difficult. It draws on areas of mathematics far removed from number theory, and introduces a new formalism that Mochizuki has named inter-universal Teichmüller theory. Even mathematicians with deep expertise in related problems have been baffled by the dense and unfamiliar arguments and the novel vocabulary ("Hodge theaters," "Frobenioids"). Jordan Ellenberg of the University of Wisconsin commented, "Looking at it, you feel a bit like you might be reading a paper from the future, or from outer space."

Yet Mochizuki's proof is being taken seriously by the mathematical community. However opaque the discourse, a few dozen determined readers are pressing on. They gather at workshops to share what they have learned; I was a spectator at one of those meetings in 2016. At some moments it reminded me of a study group formed by students struggling with a difficult course, except that the participants included distinguished senior mathematicians. I was also reminded of the spontaneous gatherings that form when nature hands us a major surprise—supernova 1987a, the discovery of crystals with fivefold symmetry. But in this case the revelation came not from nature but from a reclusive Japanese mathematician.

By the end of the weekend there were expressions of wary optimism—hope that the structure of the proof is sound, and confidence that it will eventually be understood and assimilated into the body of known mathematics. Personally, though, two days of lectures on inter-universal Teichmüller theory did not leave me with much confidence that *I* will ever understand.

Poof

These incidents and others like them have led to talk of a crisis in mathematics, and to fears that proof cannot be trusted to lead us to eternal and indubitable truth. Already in 1972 Philip J. Davis of Brown University wrote,

> The authenticity of a mathematical proof is not absolute but only probabilistic . . . . Proofs cannot be too long, else their probabilities go down, and they baffle the checking process. To put it another way: all really deep theorems are false (or at best unproved or unprovable). All true theorems are trivial.

A few years later, in *Mathematics: The Loss of Certainty*, Morris Kline portrayed mathematics as a teetering superstructure with flimsy timbers and a crumbling foundation; continuing with this architectural metaphor, he argued that proofs are "a façade rather than the supporting columns of the mathematical structure."

Davis and Kline both wrote as mathematical insiders—as members of the club, albeit iconoclastic ones. In contrast, John Horgan positioned himself as a defiant outsider when he wrote a *Scientific American* essay titled "The Death of Proof" in 1993. "The doubts riddling modern human thought have finally infected mathematics," he said. "Mathematicians may at last be forced to accept what many scientists and philosophers already have admitted: their assertions are, at best, only provisionally true, true until proved false."

My own position as an observer of these events is somewhere in the awkward middle ground, neither inside nor outside. I'll concede there is a kind of crisis going on, but only because the entire history of mathematics is just one crisis after another. The foundations are *always* crumbling, and the barbarians are *always* at the gate. When Appel and Haken published their computer-aided proof, it was hardly the first time that a technical innovation had stirred up controversy. In the seventeenth century, when algebraic methods began intruding into geometry, the heirs of the Euclidean tradition cried foul. (Hobbes was one of them.) At the end of the nineteenth century, when David Hilbert introduced nonconstructive proofs—saying, in effect, "I know x exists, but I can't tell you where to look for it"—there was another rebellion. Said one critic, "This is not mathematics. This is theology."

All in all, the crisis of the present moment seems mild compared with that of a century ago, when paradoxes in set theory led to Gottlob Frege's lament, "Alas, arithmetic totters." In response to that crisis, a rescue party of ambitious mathematicians, led by Hilbert, set out to rebuild the edifice of mathematics on a new foundation. Hilbert's plan was to apply the process of proof to proof itself, showing that the axioms and theorems of mathematics can never lead to a contradiction—that you can never prove both "x" and "not x." The outcome is well known. Kurt Gödel proved instead that if you insist on consistency, there are true statements you can't prove at all. You might think that such a Tower of Babel catastrophe would scatter the tribes of mathematics for generations, but mathematicians have carried on.

That some of the latest proofs from the frontiers of mathematical research are difficult and rely on novel tools seems to me utterly unexceptional. Of course the proofs are hard to digest; they were hard to create. These are solutions to problems that have stumped strong minds for decades or centuries. When Perelman's proof of the Poincaré conjecture defeats my attempts at understanding, this is a disappointment but not a surprise. If I have a worry about the state of mathematics, it's not the forbidding inaccessibility of the deepest thinkers; rather, it's my own clumsiness when I tackle perfectly humdrum problems, far from the frontiers of knowledge.

Who's on First?

Donald E. Knuth of Stanford University once appended a note to a computer program: "Beware of bugs in the above code; I have only proved it correct, not tried it." Coming from Knuth, the warning was a joke, but if you hear it from me, you should take it seriously.

Allow me to bring back Socrates and the slave boy for a little exercise in probability theory. They are arguing about sports. In a best-of-seven tournament (such as the baseball World Series), what is the probability that the contest will be decided in a four-game sweep? (See figure 13.5.) Assume that the teams are evenly matched, so that each team has a 50-50 chance of winning any single game.

Socrates: In a best-of-seven series, how many ways can a team score a clean sweep?

Boy: Just one way. You've got to win four in a row, with no losses.

Socrates: And how many ways could we form a five-game series, with four wins and a single loss?

Boy: Well, you could lose either the first, the second, the third or the fourth game, and win all the rest.

Socrates: And what about losing the fifth game?

Boy: If you win the first four, you don't play a fifth.

Socrates: So to build a five-game series, we take a four-game series and insert an additional loss at each position except the last, is that right?

Boy: I guess.

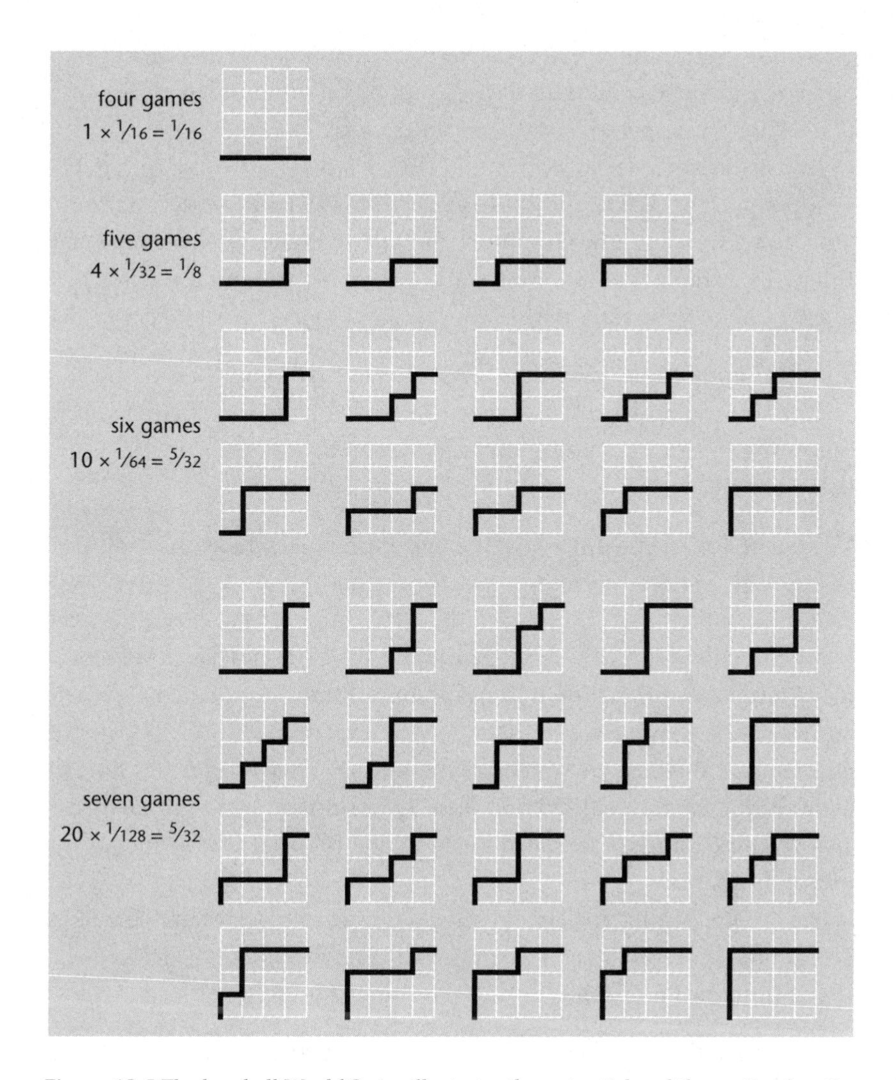

Figure 13.5 The baseball World Series illustrates the potential and the perils of mathematical reasoning. The series ends as soon as either team scores four wins. Assuming evenly matched teams, what are the probabilities that the series will last four, five, six, or seven games? Any series can be represented as a path through a 4×4grid; starting in the southwest corner, a win by one team is a step to the east, a win by the other a step to the north. The 35 cases shown are all those in which East is the series winner. If East wins each game with probability $\frac{1}{2}$, the probability of any specific five-game sequence is $(\frac{1}{2})^5$, or $\frac{1}{32}$; there are four such sequences, and so the probability that East will win in five games is $\frac{1}{8}$. Erroneous calculations either count incorrectly or assume that all cases have the same weight.

Socrates: Therefore we can create a six-game tournament by taking each of the five-game series and inserting an additional loss in each of five positions. Thus we have 4×5, or 20, six-game series.

Boy: If you say so.

Socrates: Then each of the 20 six-game series can be expanded in six different ways to make a seven-game series, and so there are 120 seven-game variations. Adding it all up, we have 1+4+20+120 cases, for a total of 145. Exactly one of these cases is a clean sweep, and so the probability is 1/145.

At this point the boy, who knows something about baseball, points out that of 105 best-of-seven World Series, 18 have been won in a clean sweep, suggesting an empirical probability nearer to $1/6$ than to $1/145$. Socrates is not swayed by this fact.

Socrates: Do not be distracted by mere appearances; here we study *ideal* baseball. Let me demonstrate by another method.

Boy: Go for it.

Socrates: Suppose for the moment that the teams always play a full schedule of seven games. Since each game can have either of two outcomes, there are 2^7, or 128, possible sequences. Now go through the list of 128 patterns and remove all those in which play continues after a team has already achieved four victories. I find that there are 70 distinct patterns remaining. Two of those patterns are clean sweeps—one for each team—and so the probability is $1/35$.

Boy: You're getting warmer. Try this: Winning four games in a row has a probability of $1/2 \times 1/2 \times 1/2 \times 1/2$, or $1/16$. Either team can have the clean sweep, so the combined probability is $1/16 + 1/16 = 1/8$.

Deductive logic has led us to three different answers, at least two of which must be wrong. Although I have made Socrates the fool in this comedy of errors, I cannot conceal that the blunders are actually my own. A few years ago I had occasion to perform this calculation, and I got a wrong answer. How did I know it was wrong? It disagreed with a simple computer experiment: a program that simulated a million random World Series and produced 124,711 four-game clean sweeps. (That's almost exactly $1/8$.)

What does it mean that I put greater trust in the output of a computer program than in my own reasoning? Well, I am *not* arguing that computer simulations are superior to proofs or more reliable than deductive methods.

It's not that there's something wrong with classical mathematics. All three approaches discussed in my pseudo-Socratic dialogue can be made to yield the right answer if they are applied with care. But it often seems that mathematical proof is foolproof only in the absence of fools.

For those who believe they could never possibly commit such an error, I offer my congratulations, along with a reminder of the infamous Monty Hall affair. In 1990 Marilyn vos Savant, a columnist for *Parade* magazine, discussed a hypothetical situation on the television game show "Let's Make a Deal," hosted by Monty Hall. A prize is hidden behind one of three doors. When a contestant chooses door 1, Hall opens door 3, showing that the prize is not there, and offers the player the option of switching to door 2. Vos Savant argued (correctly, given certain assumptions) that switching improves the odds from ⅓ to ⅔. Thousands disagreed, including more than a few mathematicians. Even Paul Erdős, a formidable probabilist, got it wrong. It was a computer simulation that ultimately persuaded him.

Stephen Stigler, eminent historian of probability and statistics, has unearthed a comparable story involving Samuel Pepys and Isaac Newton. Pepys inquired of Newton about a dice game, Is it easier to get at least one six when rolling six dice, or at least two sixes when rolling twelve dice? Newton gave the correct answer (six dice are the better bet), but he supported it with a fallacious argument, which failed to take account of the correct probability distribution. Newton also backed up his conclusion with the seventeenth-century equivalent of a computer simulation: an exact enumeration of all 6^6 or 6^{12} cases.

Putting Proof in Its Place

The law seeks proof beyond a reasonable doubt, but mathematics sets a higher standard. In a tradition that goes back to Euclid, proof is taken as a guarantee of infallibility. It is the flaming sword of a sentry standing guard over the published literature of mathematics, barring all falsehoods. And the literature may need guarding. If you view mathematics as a formal system of axioms and theorems, then the structure is dangerously brittle. Admit just one false theorem and you can prove any absurdity you please.

The special status of mathematical truth, setting the discipline apart from other arts and sciences, is a notion still cherished by many mathematicians, but proof has other roles as well; it's not just a seal of approval. David Bressoud's book *Proofs and Confirmations* gives what I believe is the

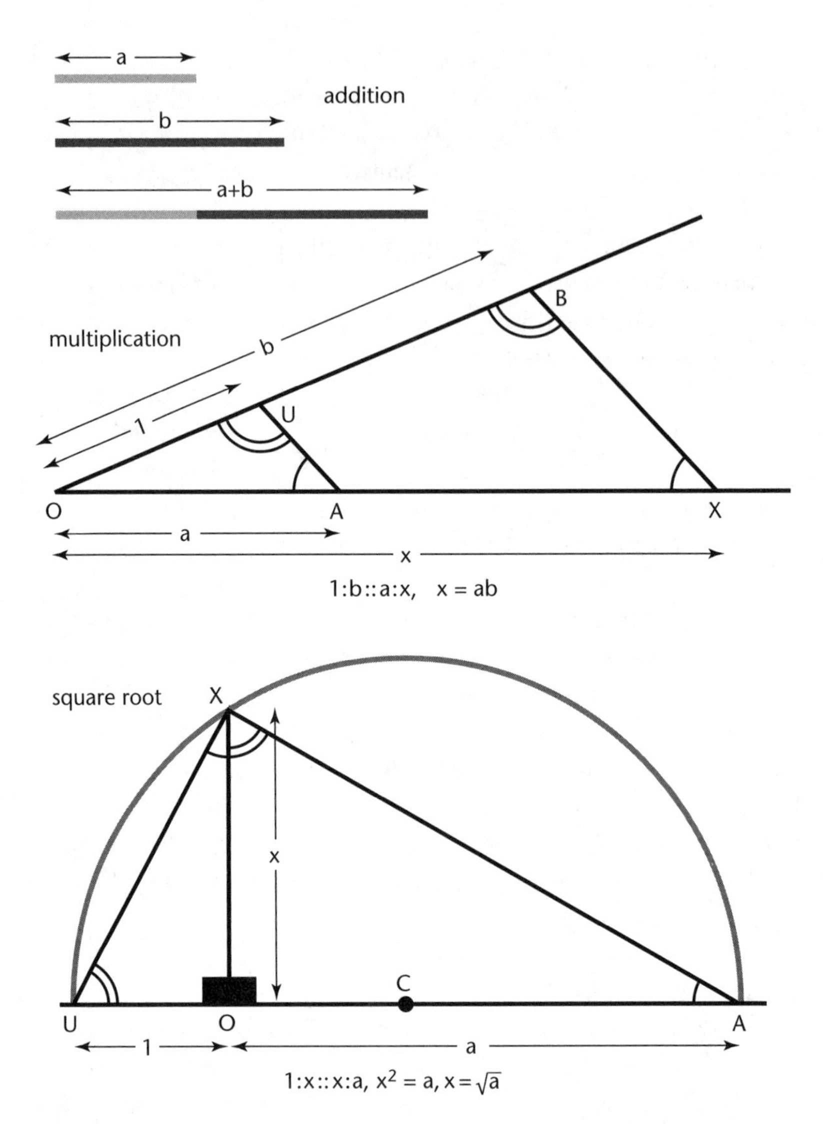

Figure 13.6 Euclidean constructions with straightedge and compass can perform five arithmetic operations: addition, subtraction, multiplication, division, and the extraction of square roots. Addition is simply the concatenation of line segments, and subtraction is the reverse. Multiplication entails building similar triangles whose sides embody the ratios 1:*b*::*a*:*x*; thus *x* is the product of *a* and *b*. Division involves reversing the process. For square roots, triangles inscribed in a semicircle yield the ratios 1:*x*::*x*:*a*, so that $x^2 = a$ and the length of the segment *x* is the square root of *a*. Only these five operations are possible; there is no way to extract cube roots, a fact crucial to proving that trisection is impossible.

best-ever insider's account of what it's like to do mathematics. Bressoud emphasizes that the most important function of proof is not to establish that a proposition is true but to explain *why* it's true: "The search for proof is the first step in the search for understanding."

And, of course, there's more to mathematics than theorems and proofs. A genre calling itself experimental mathematics is thriving today. There are journals and conferences devoted to the theme, and a series of books by Jonathan Borwein and David Bailey serve as a manifesto for the field. Not that practitioners of experimental math want to abandon or abolish proof, but they give greater scope to other activities: playing with examples, making conjectures, computation, visualization.

Still, there are ideas that never could have entered the human mind except through the reasoning process we call proof. Which brings me back to the trisection of angles.

Wantzel's Theorem

The fact that trisection is impossible is common knowledge, but the reason it's impossible—the content of the proof—is not so widely known. Many authors mention it but few explain it. Even Underwood Dudley's splendid send-up of mathematical cranks, *A Budget of Trisections*, does not go through the proof step-by-step.

The origin and history of the proof are somewhat shadowy. There is a lesson here for those who seek immortality by solving some trophy problem in mathematics. Trisection had been near the top of the most-wanted list for two millennia, and yet the author of the first published proof of impossibility has not earned a place in the pantheon.

That author was the French mathematician Pierre Laurent Wantzel (1814–1848), who is hardly a household name, even in mathematical households. His proof appeared in 1837 (when he was 23 years old). As far as I know, the paper has never been reprinted and has never been published in English translation. (I have posted my own rough translation at http://bit-player.org/extras/trisection/Wantzel1837english.pdf.) Many citations of the paper give the wrong volume number, suggesting that even some of those who refer to the proof have not read it.

A 2009 article Jesper Lützen of the University of Copenhagen asks why Wantzel's work has remained so obscure for so long. Lützen suggests several contributing factors, including Wantzel's early death; he left behind only a small and scattered body of work. But the most intriguing possibility is that Wantzel's contemporaries may have considered the proof superfluous; they believed that informal arguments presented 200 years earlier by René Descartes had settled the matter.

For modern readers, another reason for Wantzel's obscurity is that his proof is almost unintelligible. Later expositors offer more lucid accounts. Felix Klein, L. E. Dickson, and Robert C. Yates all published versions of the proof; so did Willard Van Orman Quine, who wrote his in response to a $100 challenge. For a thorough and accessible book-length exposition I highly recommend *Abstract Algebra and Famous Impossibilities,* by Arthur Jones, Sidney A. Morris, and Kenneth R. Pearson.

As penance for my youthful career as a trisector, I would now like to try giving my own brief sketch of the impossibility proof. The basic question is, What can you do with a straightedge and compass? You can draw lines and circles, obviously, but it turns out you can also do arithmetic (see figure 13.6). If the length of a line segment represents a number, then ruler-and-compass manipulations can add, subtract, multiply, and divide such numbers, and also extract square roots. Suppose you are given a segment of length 1 to start with; what further numbers can you generate? All the integers are easy to reach; you can also get to any rational number (a ratio of integers). Square roots give access to certain irrationals; by taking square roots of square roots, you can also do fourth roots, eighth roots, and so on. But that's *all* you can do. There is no way to extract cube roots, fifth roots, or any other roots that are not a power of 2—which is the crucial issue for trisection.

A purported trisection procedure is required to take an angle θ and produce $\theta/3$. Since the procedure has to work with *any* angle, we can refute it by exhibiting just one angle that cannot be trisected. The standard example is 60 degrees. Suppose the vertex of a 60-degree angle is at the origin, and one side corresponds to the positive x axis. Then to trisect the angle you must draw a line inclined by 20 degrees to the x axis and passing through the origin.

To draw any line, all you need are two points lying on the line. In this case you already have one point, namely, the origin. Thus the entire task of trisection reduces to finding one more point lying somewhere along the 20-degree line. Surely that must be easy! After all, there are infinitely many points on the line and you only need one of them. But the proof says it can't be done.

To see the source of the difficulty we can turn to trigonometry. If we knew the sine and cosine of 20 degrees, the problem would be solved; we could simply construct the point $x = \cos 20$, $y = \sin 20$, which must lie on the line. (Of course, we need the exact values; approximations from a calculator or a trig table won't help.) We *do* know the exact sine and cosine of 60 degrees; the values are $\sqrt{3}/2$ and $1/2$. Both these numbers can be constructed with ruler and compass. Furthermore, formulas relate the sine and cosine of any angle θ to the corresponding values for $\theta/3$. The formulas yield the following equation (where for brevity the symbol u replaces the expression $\cos \theta/3$):

$$\cos \theta = 4u^3 - 3u.$$

For the 60-degree angle, with $\cos \theta = 1/2$, the equation becomes $8u^3 - 6u = 1$. Note that this is a cubic equation. That's the nub of the problem. No process of adding, subtracting, multiplying, dividing, or taking square roots will ever solve the equation for the value of u. (The hard part of the proof, which I'm not brave enough to attempt here, shows that the cubic equation cannot be reduced to one of lower degree.)

This proof, with its excursions into trigonometry and algebra, would have been alien to Euclid, but the conclusion is easily translated back into the language of geometry. Not a single point along the 20-degree line (except the origin) can be reached from the 60-degree line by ruler-and-compass methods. There is something hauntingly counterintuitive about this fact. The two lines live on the same plane, they even intersect, and yet they don't communicate. You can't get there from here.

Wantzel's presentation of these ideas is not especially clever or elegant. The proof is not a good candidate for The Book, the compendium where God records the best proofs of all theorems, according to Erdős. And yet Wantzel's argument has the strength, the moxie, the bravado, that comes with the very idea of irrefutable proof. Like Hobbes, I wouldn't believe it, but the proof *compels* belief. I wonder if my old friend Dmytro would be convinced.

Sources and Resources

All the essays in this collection began as "Computing Science" columns published in *American Scientist*, the magazine of Sigma Xi, the Scientific Research Society. I am grateful to the editors and the entire magazine staff for their important contributions to the original publications.

For this volume the essays have been extensively revised and updated. I have corrected errors (many of them brought to my attention by readers), reported recent results, created new illustrations, and incorporated additional material. The original versions of the essays and related works are listed here and at http://bit-player.org/foolproof.

"Young Gauss Sums It Up" is based on "Gauss's Day of Reckoning," *American Scientist*, vol. 94, no. 3, May-June 2006, pp. 200–205. Much new material has come to light in the decade since the earlier publication, fundamentally changing the story of how Gauss's anecdote was transmitted in the years after his death. An archive of related documents, including a chronology, a bibliography, and excerpts from more than 150 tellings of the tale, is at http://bit-player.org/gauss-links.

"Outside the Law of Averages" was published in an earlier version as "Fat Tails," *American Scientist*, vol. 95, no. 3, May-June 2007, pp. 200–204. Two related blog posts discuss errors in the original article and extensions of my analysis: they are at http://bit-player.org/2007/factoidals-facts and http://bit-player.org/2007/more-factoidal-facts.

An earlier version of "How to Avoid Yourself" appeared in *American Scientist*, vol. 86, no. 4, July-August 1998, pp. 314–319. The revised essay includes new results from the past 20 years as well as better explanations of how those results were obtained.

"The Spectrum of Riemannium" first appeared in *American Scientist*, vol. 91, no. 4, July-August 2003, pp. 296–300. That version was also translated in French as "Le spectre du Riemannium" in *Pour la science*, no. 312, October 2003, and in Spanish as "El espectro di Riemannio" in *Investigacion y Ciencia*, January 2004, pp. 14–18.

"Unwed Numbers" first appeared in *American Scientist*, vol. 94, no. 1, January-February 2006, pp. 12–15. The new essay incorporates findings mentioned in "How Many Sudokus?" at http://bit-player.org/2008/how-many-sudokus, and "An Early Crop of Sudoku" at http://bit-player.org/2006/a-new-crop-of-sudoku.

"Crinkly Curves" was first published in *American Scientist*, vol. 101, no. 3, May-June 2013, pp. 178–183. It was reprinted in *The Best Writing on Mathematics 2014*, edited by Mircea Pitici (Princeton University Press, 2014). Some follow-up commentary appears in "Mapping the Hilbert Curve," at http://bit-player.org/2013/mapping-the-hilbert-curve. In addition, my interactive programs for growing and analyzing the Hilbert curve are at http://bit-player.org/extras/hilbert.

"Wagering with Zeno" first appeared in *American Scientist*, vol. 96, no. 3, May-June 2008, pp. 194–199. The present essay also draws on ideas introduced in an earlier "Computing Science" column, "Follow the Money," *American Scientist*, vol. 90, no. 5, September-October 2002, page 400–405, and in a blog post, "In Zeno's Footsteps," at http://bit-player.org/2008/in-zenos-footsteps.

"The Higher Arithmetic" was first published in *American Scientist*, vol. 97, no. 5, September-October 2009, pp. 364–368. It was reprinted in *The Best Writing on Mathematics 2010*, edited by Mircea Pitici, (Princeton University Press, 2011). An addendum at http://bit-player.org/2009/outnumbered includes some afterthoughts; an extended bibliography is at http://bit-player.org/wp-content/uploads/2009/08/higher-arithmetic-biblio.html.

"First Links in the Markov Chain" was first published in *American Scientist*, vol. 101, no. 2, March-April 2013, pp. 92–97. A supplementary blog post was published under the title "Driveling" at http://bit-player.org/2013/driveling. I have also posted a "drivel generator" showing a Markov chain in action at http://bit-player.org/wp-content/extras/drivel/drivel.html. Some of the material in this essay was also presented at a Harvard symposium, held on the occasion of the 100th anniversary of Markov's publication; for more information see http://bit-player.org/2013/100-years-of-markov-chains.

"Playing Ball in the nth Dimension" was first published as "An Adventure in the Nth Dimension," *American Scientist*, vol. 99, no. 6, November-December 2011, pp. 442–446. The magazine version was reprinted in *The Best Writing on Mathematics* 2012, edited by Mircea Pitici (Princeton University Press, 2012). Some additional thoughts were published in "The N-ball Game," at http://bit-player.org/2011/the-n-ball-game.

"Quasirandom Ramblings" was published in *American Scientist*, vol. 99, no. 4, July-August 2011, pp. 282–287. A French translation was published as "Excursions quasi-aléatoires," *Pour la Science,* no. 410, December 2011, pp. 54–60, and a German translation as "Spiel mit dem Zufall," *Spektrum der Wissenschaft,* February 2012, pp. 88–94. Additional thoughts on quasirandomness were published in "A Slight Discrepancy," at http://bit-player.org/2011/a-slight-discrepancy. For information on a Harvard talk presenting much of the same material, see http://bit-player.org/2015/a-quasirandom-talk.

"Pencil, Paper, and Pi" first appeared in *American Scientist,* vol. 102, no. 5, September-October 2014, pp. 342–345. A further discussion, with emphasis on where William Shanks went astray, appears in "The Pi Man," at http://bit-player.org/2014/the-pi-man).

The first version of "Foolproof" was published in *American Scientist,* vol. 95, no. 1, January-February 2007, pp. 10–15. A Dutch translation was published in *Nieuw Archief voor Wiskunde 5/8,* no. 2, June 2007. My translation of Wantzel's proof of the impossibility of trisecting an angle with ruler and compass is available at http://bit-player.org/wp-content/extras/trisection/Wantzel1837english.pdf.

References

Preface

Galileo. 1623. *Il Saggiatore* [The Assayer]. Rome: Giacomo Mascardi.

Russell, Bertrand. 1935. "Useless" knowledge. *In Praise of Idleness and Other Essays.* London: Allen and Unwin.

Chapter 1: Young Gauss Sums It Up

Ahrens, W. 1920. *Mathematiker-Anekdoten.* 2nd ed. Leipzig: B. G. Teubner. https://catalog.hathitrust.org/api/volumes/oclc/12406155.html.

Bell, E. T. 1937. *Men of Mathematics.* New York: Simon and Schuster.

Bieberbach, Ludwig. 1938. *Carl Friedrich Gauß: Ein Deutsches Gelehrtenleben.* Berlin: Keil.

Bühler, W. K. 1981. *Gauss: A Biographical Study.* New York: Springer.

Dunnington, G. Waldo. 1955. *Carl Friedrich Gauss: Titan of Science.* New York: Hafner. Reprinted with additional material by Jeremy Gray and Fritz-Egbert Dohse. Washington, D.C.: Mathematical Association of America, 2004.

Hall, Tord. 1970. *Carl Friedrich Gauss: A Biography,* trans. Albert Froderberg. Cambridge, Mass.: MIT Press.

Hänselmann, Ludwig. 1878. *Karl Friedrich Gauß: Zwölf Kapitel aus Seinem Leben.* Leipzig: Duncker and Humblot.

Lietzmann, W. 1918. *Riesen und Zweige im Zahlenreich* [Giants and Dwarfs in Numberland]. Leipzig: B. G. Teubner. https://archive.org/details/bub_gb_xho7AQAAIAAJ.

Mathé, Franz. 1906. Karl Friedrich Gauss. Leipzig: Weicher. https://catalog.hathitrust.org/Record/011592764.

Möbius, Paul Julius. 1899. Ueber die Anlage zur Mathematik. In *Neurologhisches Zentralblatt*, 15 November 1899, no. 22, pp. 1049–1058.

Peterson, Ivars. 2004. Young Gauss. *Science News*. https://www.sciencenews.org/article/young-gauss.

Reich, Karin. 1977. *Carl Friedrich Gauss: 1777–1977*, trans. Patricia Crampton. Bonn–Bad Godesberg: Inter Nationes.

Sartorius von Waltershausen, W. 1856. *Gauss: zum Gedächtnis*. Leipzig: S. Hirzel. English translation, *Carl Friedrich Gauss: A Memorial*, trans. Helen Worthington Gauss. Colorado Springs, Colo., 1966.

Tent, M. B. W. 2006. *The Prince of Mathematics: Carl Friedrich Gauss*. Wellesley, Mass.: A K Peters.

Winnecke, F. T. 1877. *Gauss: Ein Umriss seines Lebens und Wirkens* [Gauss: An Outline of His Life and Work]. Braunschweig, Germany: F. Vieweg. https://archive.org/details/gausseinumrisss00winngoog.

Worbs, Erich. 1955. *Carl Friedrich Gauss: Ein Lebensbild*. Leipzig: Koehler and Amelang. https://catalog.hathitrust.org/Record/000166781.

Wussing, Hans. 1989. *Carl Friedrich Gauss*. Leipzig: B. G. Teubner.

Chapter 2: Outside the Law of Averages

Adler, Robert J., Raisa E. Feldman, and Murad S. Taqqu, eds. 1998. *A Practical Guide to Heavy Tails: Statistical Techniques and Applications*. Boston: Birkhäuser.

Cannell, John Jacob. 1988. Nationally normed elementary achievement testing in America's public schools: How all 50 states are above the national average. *Educational Measurement: Issues and Practice* 7(2):5–9.

De Morgan, Augustus. 1842. *The Penny Cyclopaedia of the Society for the Diffusion of Useful Knowledge*. Vol. 23, p. 444. London: Charles Knight.

Gabaix, Xavier. 1999. Zipf's law for cities: An explanation. *Quarterly Journal of Economics* 114(3):739–767.

Galambos, Janos, and Italo Simonelli. 2004. *Products of Random Variables: Applications to Problems of Physics and to Arithmetical Functions*. New York: Marcel Dekker.

Keillor, Garrison. 1985. *Lake Wobegon Days*. New York: Viking.

Kramp, Christian. 1808. Preface. In *Elements d'arithmétique universelle*. Cologne.

Mitzenmacher, Michael. 2004. A brief history of generative models for power law and lognormal distributions. *Internet Mathematics* 1(2):226–251.

Newman, M. E. J. 2005. Power laws, Pareto distributions and Zipf's law. *Contemporary Physics* 46:323–351.

Pakes, Anthony G. 2008. Tails of stopped random products: The factoid and some relatives. *Journal of Applied Probability* 45:1161–1180.

Reed, William J., and Barry D. Hughes. 2002. From gene families and genera to incomes and internet file sizes: Why power laws are so common in nature. *Physical Review E* 66(6):067103.

Sigman, Karl. 1999. A primer on heavy-tailed distributions. *Queueing Systems: Theory and Applications* 33(1–3):261–275.

Chapter 3: How to Avoid Yourself

Alexandrowicz, Z. 1969. Monte Carlo of chains with excluded volume: A way to evade sample attrition. *Journal of Chemical Physics* 51:561–565.

Conway, A. R., I. G. Enting, and A. J. Guttmann. 1993. Algebraic techniques for enumerating self-avoiding walks on the square lattice. *Journal of Physics A: Mathematical and General* 26:1519–1534.

Conway, A. R., and A. J. Guttmann. 1996. Square lattice self-avoiding walks and corrections to scaling. *Physical Review Letters* 77:5284–5287.

Domb, C., and M. E. Fisher. 1958. On random walks with restricted reversals. *Proceedings of the Cambridge Philosophical Society* 54:48–59.

Duminil-Copin, H., and S. Smirnov. 2012. The connective constant of the honeycomb lattice equals $\sqrt{2+\sqrt{2}}$. *Annals of Mathematics* 175(3):1653–1665.

Enting, Ian G. 1980. Generating functions for enumerating self-avoiding rings on the square lattice. *Journal of Physics A: Mathematical and General* 13:3713–3722.

Fisher, Michael E., and M. F. Sykes. 1959. Excluded-volume problem and the Ising model of ferromagnetism. *Physical Review* 114:45–58.

Guttmann, A. J., T. Prellberg, and A. L. Owczarek. 1993. On the symmetry classes of planar self-avoiding walks. *Journal of Physics A: Mathematical and General* 26:6615–6623.

Guttmann, A. J., and Jian Wang. 1991. The extension of self-avoiding random walk series in two dimensions. *Journal of Physics A: Mathematical and General* 24:3107–3109.

Hayes, Brian. 1998. Prototeins. *American Scientist* 86:216–221.

Hughes, Barry D. 1995. *Random Walks and Random Environments*. Vol. 1: *Random Walks*. Oxford: Clarendon Press.

Jacobsen, J. L., C. R. Scullard, and A. J. Guttmann. 2016. On the growth constant for square-lattice self-avoiding walks. https://arxiv.org/abs/1607.02984.

Jensen, I. 1994. Enumeration of self-avoiding walks on the square lattice. *Journal of Physics A: Mathematical and General* 37(21):5503–5524.

———. 2013. A new transfer-matrix algorithm for exact enumerations: self-avoiding walks on the square lattice. https://arxiv.org/abs/1309.6709.

Lal, Moti. 1969. "Monte Carlo" computer simulation of chain molecules, I. *Molecular Physics* 17:57–64.

Madras, Neal, and Gordon Slade. 1993. *The Self-Avoiding Walk*. Boston: Birkhäuser.

Madras, Neal, and Alan D. Sokal. 1988. The pivot algorithm: A highly efficient Monte Carlo method for the self-avoiding walk. *Journal of Statistical Physics* 50:109–186.

Nienhuis, Bernard. 1982. Exact critical point and exponents of the O(n) model in two dimensions. *Physical Review Letters* 49:1062.

O'Brien, George L. 1990. Monotonicity of the number of self-avoiding walks. *Journal of Statistical Physics* 59:969–979.

Pólya G. 1921. Über eine Aufgabe der Wahrscheinlichkeitsrechnung betreffend die Irrfahrt im Strassennetz. *Mathematische Annalen* 84(1-2):149–160. http://eudml.org/doc/158886.

Slade, Gordon. 1994. Self-avoiding walks. *Mathematical Intelligencer* 16(1):29–35.

———. 1996. Random walks. *American Scientist* 84:146–153.

Sykes, M. F. 1961. Some counting theorems in the theory of the Ising model and the excluded volume problem. *Journal of Mathematical Physics* 2:52–62.

Wang, Jian. 1989. A new algorithm to enumerate the self-avoiding random walk. *Journal of Physics A: Mathematical and General* 22:L969–L971.

Chapter 4: The Spectrum of Riemannium

Aldous, David, and Persi Diaconis. 1999. Longest increasing subsequences: From patience sorting to the Baik-Deift-Johansson theorem. *Bulletin of the American Mathematical Society* 36:413–432.

Baik, Jinho, Alexei Borodin, Percy Deift, and Toufic Suidan. 2006. A model for the bus system in Cuernavaca (Mexico). *Journal of Physics A: Mathematical and General* 39(28):8965–8975.

Baik, Jinho, Percy Deift, and Kurt Johansson. 1999. On the distribution of the length of the longest increasing subsequence of random permutations. *Journal of the American Mathematical Society* 12:1119–1178.

Bohigas, Oriol. 2005. Compound nucleus resonances, random matrices, quantum chaos. In *Recent Perspectives in Random Matrix Theory and Number Theory*, ed. F. Mezzadri and N. C. Snaith. Cambridge: Cambridge University Press.

Bohigas, Oriol, and Marie-Joya Giannoni. 1984. Chaotic motion and random matrix theories. In *Mathematical and Computational Methods in Nuclear Physics*, ed. J. S. Dehesa, J. M. G. Gomez, and A. Polls, pp. 1–99. New York: Springer.

Diaconis, Persi. 2003. Patterns in eigenvalues: The 70th Josiah Willard Gibbs lecture. *Bulletin of the American Mathematical Society* 40:155–178.

Dotsenko, Victor. 2011. Universal randomness. *Physics-Uspekhi* 54(3):259–280.

Dyson, Freeman J. 1962. Statistical theory of energy levels of complex systems, I, II, and III. *Journal of Mathematical Physics* 3:140–175.

Firk, Frank W. K., and Steven J. Miller. 2009. Nuclei, primes, and the random matrix connection. *Symmetry* 1(1):64–105.

Forrester, P. J., N. C. Snaith, and J. J. M. Verbaarschot. 2003. Developments in random matrix theory. Introduction to a special issue on random matrix theory. *Journal of Physics A: Mathematical and General* 36(12):R1–R10.

Hiary, G. A., and A. M. Odlyzko. 2012. The zeta function on the critical line: Numerical evidence for moments and random matrix theory models. *Mathematics of Computation* 81(279):1723–1752.

Katz, Nicholas M., and Peter Sarnak. 1999. Zeroes of zeta functions and symmetry. *Bulletin of the American Mathematical Society* 36:1–26.

Krbálek, Milan, and Petr Šeba. 2000. The statistical properties of the city transport in Cuernavaca (Mexico) and random matrix ensembles. *Journal of Physics A: Mathematical and General* 33(26):L229–L234.

LeBoeuf, P., A. G. Monastra, and O. Bohigas. 2001. The Riemannium. *Regular and Chaotic Dynamics* 6(2):205–210.

Liou, H. I., H. S. Camarda, S. Wynchank, M. Slagowitz, G. Hacken, F. Rahn, and J. Rainwater. 1972. Neutron resonance spectroscopy. VIII: The separated isotopes of erbium: Evidence for Dyson's theory concerning level spacings. *Physical Review C* 5:974–1001.

Montgomery, H. L. 1973. The pair correlation of zeros of the zeta function. *Proceedings of Symposia in Pure Mathematics* 24:181–193.

Odlyzko, A. 1992. The 10^{20}th zero of the Riemann zeta function and 175 million of its neighbors. Unpublished manuscript. http://www.dtc.umn.edu/~odlyzko/unpublished/index.html.

Riemann, Bernhard. 1859. Über die Anzahl der Primzahlen unter einer gegebenen Grösse [On the number of primes less than a given magnitude]. *Monatsberichte der Berliner Akademie*. Collected in *Gesammelte Werke*, 1892. Leipzig: Teubner. Reissued as *The Collected Works of Bernhard Riemann*, ed. H. Weber, 1953. New York: Dover.

Terras, Audrey. 2002. Finite quantum chaos. *American Mathematical Monthly* 109:121–139

Wigner, Eugene P. 1957. Statistical properties of real symmetric matrices with many dimensions. *Canadian Mathematical Congress Proceedings*, pp. 174–184. Reprinted in *Statistical Theories of Spectra: Fluctuations. A Collection of Reprints and Original Papers, with an Introductory Review*, Charles E. Porter, ed, 1965, New York and London: Academic Press, pp. 188–198.

Wolchover, Natalie. 2014. At the far ends of a new universal law. *Quanta*. https://www.quantamagazine.org/20141015-at-the-far-ends-of-a-new-universal-law/.

Chapter 5: Unwed Numbers

Bailey, R. A., Peter J. Cameron, and Robert Connelly. 2008. Sudoku, gerechte designs, resolutions, affine space, spreads, reguli and Hamming codes. *American Mathematical Monthly* 115:383–404.

Behrens, W. U. 1956. Feldversuchsanordnungen mit verbessertem Ausgleich der Bodenunterschiede. *Zeitschrift für Landwirtschaftliches Versuchs- und Untersuchungs wesen* 2:176–193.

Boyer, Christian. 2006. Les ancêtres français du sudoku. *Pour la Science* 344:8–11.

Chen, Zhe. 2009. Heuristic reasoning on graph and game complexity of Sudoku. https://arxiv.org/abs/0903.1659.

Crook, J. F. 2009. A pencil-and-paper algorithm for solving Sudoku puzzles. *Notices of the American Mathematical Society* 54(4):460–468.

Davis, Tom. 2012. The mathematics of Sudoku. http://www.geometer.org/mathcircles/sudoku.pdf.

Eppstein, David. 2005. Nonrepetitive paths and cycles in graphs with application to Sudoku. http://arxiv.org/abs/cs.DS/0507053.

Felgenhauer, Bertram, and Frazer Jarvis. 2005. Enumerating possible Sudoku grids. (See Sudoku enumeration problems. http://www.afjarvis.staff.shef.ac.uk/sudoku/.)

McGuire, Gary, Bastian Tugemann, and Gilles Civario. 2012. There is no 16-clue Sudoku: Solving the Sudoku minimum number of clues problem. https://arxiv.org/abs/1201.0749.

Royle, Gordon. Minimum Sudoku. http://staffhome.ecm.uwa.edu.au/~00013890/sudokumin.php.

Russell, Ed, and Frazer Jarvis. 2006. Mathematics of Sudoku II. *Mathematical Spectrum* 39:54–58.

Yato, Takayuki, and Takahiro Seta. 2002. Complexity and completeness of finding another solution and its application to puzzles. *Information Processing Society of Japan SIG Notes* 2002-AL-87-2.

Chapter 6: Crinkly Curves

Bader, M. 2013. *Space-Filling Curves: An Introduction with Applications in Scientific Computing*. Berlin: Springer.

Bader, M., and C. Zenger. 2006. Cache oblivious matrix multiplication using an element ordering based on a Peano curve. *Linear Algebra and Its Applications* 417(2-3):301–313.

Bartholdi, J. J. III, L. K. Platzman, R. L. Collins, and W. H. Warden III. 1983. A minimal technology routing system for Meals on Wheels. *Interfaces* 13(3):1–8.

Dauben, J. W. 1979. *Georg Cantor: His Mathematics and Philosophy of the Infinite*. Cambridge, Mass.: Harvard University Press.

Gardner, M. 1976. Mathematical games: In which "monster" curves force redefinition of the word "curve." *Scientific American* 235:124–133.

Haverkort, Herman. 2016. How many three-dimensional Hilbert curves are there? https://arxiv.org/abs/1610.00155.

Hilbert, D. 1891. Über die stetige Abbildung einer Linie auf ein Flächenstück. *Mathematische Annalen* 38:459–460.

———. 1926. Über das Unendliche [On the infinite]. *Mathematische Annalen* 95:161–190.

Moore, E. H. 1900. On certain crinkly curves. *Transactions of the American Mathematical Society* 1(1):72–90.

Null, A. 1971. Space-filling curves, or how to waste time with a plotter. *Software: Practice and Experience* 1:403–410.

Peano, G. 1890. Sur une courbe, qui remplit toute une aire plane. *Mathematische Annalen* 36:157–160.

Platzman, L. K., and J. J. Bartholdi III. 1989. Spacefilling curves and the planar travelling salesman problem. *Journal of the Association for Computing Machinery* 36:719–737.

Sagan, H. 1991. Some reflections on the emergence of space-filling curves: The way it could have happened and should have happened, but did not happen. *Journal of the Franklin Institute* 328:419–430.

———. 1994. *Space-Filling Curves*. New York: Springer.

Sierpiński, W. 1912. Sur une nouvelle courbe continue qui remplit toute une aire plane. *Bulletin de l'Académie des Sciences de Cracovie*, Série A, 462–478.

Velho, L., and J. de Miranda Gomes. 1991. Digital halftoning with space filling curves. *Computer Graphics* 25(4):81–90.

Chapter 7: Wagering with Zeno

Diaconis, P. 1988. Recent progress on de Finetti's notions of exchangeability. *Bayesian Statistics 3*, J. M. Bernardo, M. H. DeGroot, D. V. Lindley, and A. F. M. Smith, eds., pp. 111–125,

Freedman, David A. 1965. Bernard Friedman's urn. *Annals of Mathematical Statistics* 36:956–970.

Friedman, Bernard. 1949. A simple urn model. *Communications on Pure and Applied Mathematics* 2:59–70.

Hayes, Brian. 2002. Follow the money. *American Scientist* 90(5):400–405.

Johnson, Norman L., and Samuel Kotz. 1977. *Urn Models and Their Application: An Approach to Modern Discrete Probability Theory*. New York: Wiley.

Krapivsky, P. L., and S. Redner. 2004. Random walk with shrinking steps. *American Journal of Physics* 72:591–598.

Pemantle, Robin. 2007. A survey of random processes with reinforcement. *Probability Surveys* 4:1–79.

Steinsaltz, David. 1997. Zeno's walk: A random walk with refinements. *Probability Theory and Related Fields* 107:99–121.

Chapter 8: The Higher Arithmetic

Bailey, David H. 2005. High-precision floating-point arithmetic in scientific computation. *Computing in Science and Engineering* 7(3):54–61.

Clenshaw, C. W., and F. W. J. Olver. 1984. Beyond floating point. *Journal of the Association for Computing Machinery* 31:319–328.

Clenshaw, C. W., F. W. J. Olver, and P. R. Turner. 1989. Level-index arithmetic: An introductory survey. In *Numerical Analysis and Parallel Processing: Lectures Given at the Lancaster Numerical Analysis Summer School, 1987*, pp. 95–168. Berlin: Springer.

Feldstein, Alan, and Peter R. Turner. 2006. Gradual and tapered overflow and underflow: A functional differential equation and its approximation. *Journal of Applied Numerical Mathematics* 56(3):517–532.

Feynman, R. 1987. Guest lecture at California Institute of Technology. Quoted in Richard P. Feynman, Teacher, by David Goodstein, *Physics Today* 42, no. 2 (1989):93.

Goldberg, David. 1991. What every computer scientist should know about floating-point arithmetic. *ACM Computing Surveys* 23(1):5–48.

Gustafson, John L. 2015. *The End of Error: Unum Computing*. Boca Raton, Fla.: CRC Press.

Hamada, Hozumi. 1987. A new real number representation and its operation. In *Proceedings of the Eighth Symposium on Computer Arithmetic*, pp. 153–157. Washington, D.C.: IEEE Computer Society Press.

IEEE Computer Society. 2008. IEEE Std-754-2008 Standard for Floating Point Arithmetic.

Lozier, Daniel W. 1993. An underflow-induced graphics failure solved by SLI arithmetic. In *Proceedings of the 11th Symposium on Computer Arithmetic*, pp. 10–17. Los Alamitos, Calif.: IEEE Computer Society Press.

———, and F. W. J. Olver. 1990. Closure and precision in level-index arithmetic. *SIAM Journal on Numerical Analysis* 27:1295–1304.

Matsui, Shouichi, and Masao Iri. 1981. An overflow/underflow-free floating-point representation of numbers. *Journal of Information Processing* 4:123–133.

Morris, Robert. 1971. Tapered floating point: A new floating-point representation. *IEEE Transactions on Computers* C-20:1578–1579.

Muller, Jean-Michel, Nicolas Brisebarre, Florent de Dinechin, Claude-Pierre Jeannerod, Vincent Lefèvre, Guillaume Melquiond, Nathalie Revol, Damien Stehlé, and Serge Torres. 2010. *Handbook of Floating-Point Arithmetic*. Boston: Birkhäuser.

Turner, Peter R. 1991. Implementation and analysis of extended SLI operations. In *Proceedings of the 10th Symposium on Computer Arithmetic*, pp. 118–126. Los Alamitos, Calif.: IEEE Computer Society Press.

Chapter 9: First Links in the Markov Chain

Ash, Robert B., and Richard L. Bishop. 1972. Monopoly as a Markov process. *Mathematics Magazine* 45:26–29.

Basharin, G. P., A. N. Langville, and V. A. Naumov. 2004. The life and work of A. A. Markov. *Linear Algebra and Its Applications* 386:3–26.

Bukiet, Bruce, Elliotte Harold, and José Luis Palacios. 1997. A Markov chain approach to baseball. *Operations Research* 45(1):14–23.

Diaconis, Persi. 2009. The Markov chain Monte Carlo revolution. *Bulletin of the American Mathematical Society* 46:179–205.

Graham, Loren R., and Jean-Michel Kantor. 2009. *Naming Infinity: A True Story of Religious Mysticism and Mathematical Creativity*. Cambridge, Mass.: Belknap Press of Harvard University Press.

Kemeny, J. G., J. L. Snell, and A. W. Knapp. 1976. *Denumerable Markov Chains*. New York: Springer.

Link, D. 2006. Chains to the West: Markov's theory of connected events and its transmission to Western Europe. *Science in Context* 19(4):561–589.

———. 2006. Traces of the mouth: Andrei Andreyevich Markov's mathematization of writing. *History of Science* 44(145):321–348.

Markov, A. A. 1913. An example of statistical investigation of the text *Eugene Onegin* concerning the connection of samples in chains. (In Russian.) *Bulletin of the Imperial Academy of Sciences of St. Petersburg* 7(3):153–162. Unpublished English translation by Morris Halle, 1955. English translation by Alexander Y. Nitussov, Lioudmila Voropai, Gloria Custance, and David Link, 2006. *Science in Context* 19(4):591–600.

Ondar, Kh. O., ed. 1981. *The Correspondence Between A. A. Markov and A. A. Chuprov on the Theory of Probability and Mathematical Statistics*. New York: Springer.

Pushkin, A. S. 1833. *Eugene Onegin: A Novel in Verse,* trans. Charles Johnston. London: Penguin, 1977.

Seneta, E. 1996. Markov and the birth of chain dependence theory. *International Statistical Review* 64:255–263.

———. 2003. Statistical regularity and free will: L. A. J. Quetelet and P. A. Nekrasov. *International Statistical Review* 71:319–334.

Shannon, C. E. 1948. A mathematical theory of communication. *Bell System Technical Journal* 27:379–423, 623–656.

Sheynin, O. B. 1989. A. A. Markov's work on probability. *Archive for History of Exact Sciences* 39(4):337–377.

Vucinich, A. 1960. Mathematics in Russian culture. *Journal of the History of Ideas* 21(2):161–179.

Chapter 10: Playing Ball in the *n*th Dimension

Ball, K. 1997. An elementary introduction to modern convex geometry. In *Flavors of Geometry,* ed. Silvio Levy. Cambridge: Cambridge University Press.

Bellman, R. E. 1961. *Adaptive Control Processes: A Guided Tour.* Princeton, N.J.: Princeton University Press.

Catalan, Eugène. 1839, 1841. *Journal de Mathématiques Pures et Appliquées* 4:323–344; 6:81–84.

Cipra, B. 1991. Here's looking at Euclid. In *What's Happening in the Mathematical Sciences.* Vol. 1, p. 25. Providence: American Mathematical Society.

Clifford, W. K. 1866. Question 1878. *Mathematical Questions, with Their Solutions, from the "Educational Times"* 6:83–87.

Conway, J. H., and N. J. A. Sloane. 1999. *Sphere Packings, Lattices, and Groups.* 3rd ed. New York: Springer.

Heyl, P. R. 1897. Properties of the locus r = constant in space of n dimensions. Philadelphia: Publications of the University of Pennsylvania, Mathematics, No. 1, 1897, pp. 33–39. Available at http://books.google.com/books?id=j5pQAAAAYAAJ

On-Line Encyclopedia of Integer Sequences, sequence A074455. Published electronically at http://oeis.org, 2010.

Schläfli, L. 1858. On a multiple integral, trans. A. Cayley. *Quarterly Journal of Pure and Applied Mathematics* 2:269–301.

Sommerville, D. M. Y. 1911. *Bibliography of Non-Euclidean Geometry, Including the Theory of Parallels, the Foundation of Geometry, and Space of N Dimensions*. London: Harrison.

———. 1929. *An Introduction to the Geometry of N Dimensions*. New York: Dover Publications.

Wikipedia. 2016. Volume of an *n*-ball. https://en.wikipedia.org/wiki/Volume_of_an_n-ball.

Chapter 11: Quasirandom Ramblings

Bork, Alfred M. 1967. Randomness and the twentieth century. *Antioch Review* 27(1):40–61.

Braverman, Mark. 2011. Poly-logarithmic independence fools bounded-depth boolean circuits. *Communications of the ACM* 54(4):108–115.

Caflisch, R. E. 1998. Monte Carlo and quasi–Monte Carlo methods. *Acta Numerica* 7:1–49.

Chazelle, Bernard. 2000. *The Discrepancy Method: Randomness and Complexity*. Cambridge: Cambridge University Press.

Dyer, Martin, and Alan Frieze. 1991. Computing the volume of convex bodies: A case where randomness provably helps. In *Probabilistic Combinatorics and Its Applications*, ed. Béla Bollobás, pp. 123–169. Providence: American Mathematical Society.

Galanti, Silvio, and Alan Jung. 1997. Low-discrepancy sequences: Monte Carlo simulation of option prices. *Journal of Derivatives* 5(1):63–83.

Householder, A. S., G. E. Forsythe, and H. H. Germond, eds. 1951. *Monte Carlo Method: Proceedings of a Symposium*. National Bureau of Standards Applied Mathematics Series. Vol. 12. Washington, D.C.: Government Printing Office.

Karp, Richard M. 1991. An introduction to randomized algorithms. *Discrete Applied Mathematics* 34:165–201.

Kuipers, L., and H. Niederreiter. 1974. *Uniform Distribution of Sequences*. New York: Dover Publications.

Kuo, Frances Y., and Ian H. Sloan. 2005. Lifting the curse of dimensionality. *Notices of the American Mathematical Society* 52:1320–1328.

Matousek, Jiri. 1999, 2010. *Geometric Discrepancy: An Illustrated Guide*. Heidelberg: Springer.

Metropolis, Nicholas. 1987. The beginning of the Monte Carlo method. *Los Alamos Science* 15:125–130.

———, and S. Ulam. 1949. The Monte Carlo method. *Journal of the American Statistical Association* 247:335–341.

Motwani, Rajeev, and Prabhakar Raghavan. 1995. *Randomized Algorithms*. Cambridge: Cambridge University Press.

Niederreiter, Harald. 1978. Quasi–Monte Carlo methods and pseudo-random numbers. *Bulletin of the American Mathematical Society* 84(6):957–1041.

———. 1992. *Random Number Generation and Quasi–Monte Carlo Methods*. Philadelphia: SIAM.

Paskov, Spassimir H., and Joseph E. Traub. 1995. Faster valuation of financial derivatives. *Journal of Portfolio Management* 22(1):113–121.

Richtmyer, R. D. 1951. The evaluation of definite integrals, and a quasi–Monte Carlo method based on the properties of algebraic numbers. Report LA-1342. Los Alamos Scientific Laboratory. http://www.osti.gov/bridge/product.biblio.jsp?osti_id=4405295

Roth, K. F. 1954. On irregularities of distribution. *Mathematika: Journal of Pure and Applied Mathematics* 1:73–79.

Sloan, Ian H., and Henryk Woźniakowski. 1998. When are quasi–Monte Carlo algorithms efficient for high dimensional integrals? *Journal of Complexity* 14:1–33.

Stigler, Stephen M. 1991. Stochastic simulation in the nineteenth century. *Statistical Science* 6:89–97.

von Neumann, John. 1951. Various techniques used in connection with random digits. (Summary written by George E. Forsythe.) In *Monte Carlo Method: Proceedings of a Symposium*. National Bureau of Standards Applied Mathematics Series. Vol. 12, pp. 36–38. Washington, D.C.: Government Printing Office.

Zaremba, S. K. 1968. The mathematical basis of Monte Carlo and quasi–Monte Carlo methods. *SIAM Review* 10:303–314.

Chapter 12: Pencil, Paper, and Pi

Arndt, Jörg, and Christoph Haenel. 2001. Pi Unleashed. Translated from the German by Catriona and David Lischka. Berlin: Springer.

Berggren, Lennart, Jonathan Borwein, and Peter Borwein, eds. 2004. *Pi, A Source Book*. 3rd ed. New York: Springer.

Engert, Erwin. Undated manuscripts on pencil-and-paper calculations of pi. https://www.engert.us/erwin/Miscellaneous.html.

Ferguson, D. F. 1946. Evaluation of π. Are Shanks' figures correct? *Mathematical Gazette* 30(289):89–90.

O'Connor, J. J., and E. F. Robertson. 2007. William Shanks. In the Mactutor History of Mathematics, University of St. Andrews. http://www-history.mcs.st-andrews.ac.uk/Biographies/Shanks.html.

Rutherford, William. 1853. On the extension of the value of the ratio of the circumference of a circle to its diameter. *Proceedings of the Royal Society of London* 6:273–275.

Shanks, William. 1853. *Contributions to Mathematics, Comprising Chiefly the Rectification of the Circle to 607 Places of Decimals*. London: G. Bell.

———. 1854. On the extension of the value of the base of Napier's logarithms; of the Napierian logarithms of 2, 3, 5, and 10; and of the modulus of Briggs's, or the common system of logarithms; all to 205 places of decimals. *Proceedings of the Royal Society of London* 6:397–398.

———. 1866. On the calculation of the numerical value of Euler's constant, which Professor Price, of Oxford, calls E. *Proceedings of the Royal Society of London* 15:429–432.

———. 1873. On the extension of the numerical value of π. *Proceedings of the Royal Society of London* 21:318–319.

Smith, L. B., J. W. Wrench, and D. F. Ferguson. 1947. A new approximation to pi. *Mathematical Tables and Other Aids to Computation* 2(18):245–248.

Wrench, J. W. Jr. 1960. The evolution of extended decimal approximations to π. *Mathematics Teacher* 53(8): 644–650.

Chapter 13: Foolproof

Aigner, Martin, and Günter M. Ziegler. 1998. *Proofs from The Book*. Berlin: Springer.

Appel, K., and W. Haken. 1986. The four color proof suffices. *Mathematical Intelligencer* 8(1):10–20.

Aubrey, John. 1898. *Brief Lives, Chiefly of Contemporaries, Set Down by John Aubrey, between the Years 1669 & 1696*, ed. Andrew Clark. Oxford: Clarendon Press.

Borwein, Jonathan, and David Bailey. 2004. *Mathematics by Experiment: Plausible Reasoning in the 21st Century*. Natick, Mass.: A K Peters.

Bressoud, David M. 1999. *Proofs and Confirmations: The Story of the Alternating Sign Matrix Conjecture*. Cambridge: Cambridge University Press.

Davis, P. J. 1972. Fidelity in mathematical discourse: Is one and one really two? *American Mathematical Monthly* 79(3):252–263.

Dickson, Leonard Eugene. 1921. Why it is impossible to trisect an angle or to construct a regular polygon of 7 or 9 sides by ruler and compasses. *Mathematics Teacher* 14:217–218.

Dudley, Underwood. 1987. *A Budget of Trisections*. New York: Springer.

Ellenberg, Jordan. 2012. Mochizuki on ABC. https://quomodocumque.wordpress.com/2012/09/03/mochizuki-on-abc/.

Gardner, Martin. 1966. Mathematical games: The persistence (and futility) of efforts to trisect the angle. *Scientific American* 214(6):116–122.

Hales, Thomas C. 2005. A proof of the Kepler conjecture. *Annals of Mathematics* 162(3):1065–1185; see also special issue of *Discrete and Computational Geometry* 36, no. 1 (2006):1–265.

Horgan, John. 1993. The death of proof. *Scientific American* 269(4):92–103.

Jesseph, Douglas M. 1999. *Squaring the Circle: The War between Hobbes and Wallis*. Chicago: University of Chicago Press.

Jones, Arthur, Sidney A. Morris, and Kenneth R. Pearson. 1991. *Abstract Algebra and Famous Impossibilities*. New York: Springer.

Klein, Felix. 1897. *Famous Problems of Elementary Geometry: The Duplication of the Cube, the Trisection of an Angle, the Quadrature of the Circle*. Boston: Ginn.

Kline, Morris. 1980. *Mathematics: The Loss of Certainty*. New York: Oxford University Press.

Lützen, Jesper. 2009. Why was Wantzel overlooked for a century? The changing importance of an impossibility result. *Historia Mathematica* 36:374–394.

Mochizuki, Shinichi. 2012–2017. Inter-universal Teichmüller theory, parts I–IV (updated). http://www.kurims.kyoto-u.ac.jp/~motizuki/papers-english.html.

Quine, W. V. 1990. Elementary proof that some angles cannot be trisected by ruler and compass. *Mathematics Magazine* 63(2):95–105.

Sormani, Christina. 2003–2010. Hamilton, Perelman and the Poincaré conjecture. http://comet.lehman.cuny.edu/sormani/others/perelman/introperelman.html, and a mirror site (2017) at https://sites.google.com/site/professorsormani/home/outreach/introperelman

Stigler, Stephen. 2006. Isaac Newton as a probabilist. *Statistical Science* 21:400–403.

Wantzel, Pierre Laurent. 1837. Recherches sur les moyens de reconnaître si un problème de géométrie peut se résoudre avec la règle et le compas. *Journal de Mathématiques Pures et Appliquées* 2:366–372.

Yates, Robert C. 1942. *The Trisection Problem*. Baton Rouge, La.: Franklin Press.

Index

Page numbers in *italic type* refer to figures and their captions.